# MEAT MAKES PEOPLE POWERFUL

# MEAT MAKES PEOPLE POWERFUL

## A Global History of the Modern Era

*Wilson J. Warren*

UNIVERSITY OF IOWA PRESS, IOWA CITY

University of Iowa Press, Iowa City 52242

www.uipress.uiowa.edu
Printed in the United States of America

Design by Shaun Allshouse



The University of Iowa Press is a member of Green Press Initiative and is committed to preserving natural resources.

Printed on acid-free paper

Cataloging-in-Publication data for *Meat Makes People Powerful* by Wilson J. Warren is on file at the Library of Congress.

ISBN 978-1-60938-555-2 (pbk)
ISBN 978-1-60938-556-9 (ebk)

*For John, James, and Katherine*

# Contents

# Preface and Acknowledgments

I started this project in 2008 when I was a Fulbright lecturer in American Studies at Doshisha University in Kyoto, Japan. Although I am a modern US historian by training, I have a long-standing interest in Japan. My first experience there was as an English teacher for the Japan Exchange and Teaching (JET) program in Miyazaki in 1991 and 1992. While teaching on Doshisha's Kyoto campus, located adjacent to the imperial palace grounds, I discovered the Go'ō Shrine, also located next to the imperial grounds about a mile from the campus. Go'ō is a small Shinto shrine with an entrance guarded by wild boars instead of the usual stone-carved mythical lions, or *komainu*. The shrine commemorates Lord Wake no Kiyomaro, who lived in the eighth century. Wake no Kiyomaro was banished to Japan's far south because he opposed an attempt by female emperor Kōken (r. 749–758)—who had returned to the throne as Shōtoku (r. 764–770)—to place the Buddhist cleric Dōkyō on the throne after her. Not only was Wake no Kiyomaro banished, but his leg tendons were severed to hobble him. According to folklore, as Wake no Kiyomaro's carriage traveled in the Kagoshima countryside, he was suddenly surrounded by three hundred wild boars. Instead of attacking him, they escorted him to safety. In the process, his leg injuries were healed. The shrine is therefore popular among those with leg injuries.

Although Lord Wake no Kiyomaro is revered as a deity in the Go'ō Shrine, the grounds include an amazing array of displays featuring wild boars. Boars have powerful totem meaning in many cultures, including East Asia, and they are also one of the few wild animals that still roam parts of Japan. The Go'ō Shrine includes statues, paintings, and other depictions of boars, including stuffed ones. My daughter and I bought boar good luck charms, but not because of concerns about leg injuries. Although the story of Lord Wake no Kiyomara has nothing to do with eating wild boar, the shrine is thought-provoking about the role of human-animal relations in Japan's history. Because I was working on Japan's meat history, each time I visited there I thought about its people's long history of avoiding many types of animal foods and then how rapidly Japanese people had incorporated meat into their diets in the modern era. As part of a longer parody on Japan's rapid Westernization, Kanagaki Robun, one of the country's Meiji era meat skeptics, suggested that beef eating makes people powerful. Borrowing from Kanagaki, I find that *Meat Makes People Powerful* is useful as a book title because it suggests the various reasons why people have incorporated and desired animal products. I hope this book sheds some light on the multiple dimensions and effects that accompanied meat's history in Japan as well as other nations around the globe. Please note that this text follows Japanese custom and scholarly convention by noting Japanese names with surname first and given name last.

I have been working on projects related to meat history for over three decades and have been helped by many people during that time. For this project, I would like to acknowledge the Japanese translation assistance I received from several Doshisha University American studies graduate students: Hara Emiko, Izumi Masashi, Komatsu Tobihiko, and Nishijima Yumi. My American studies colleagues at Doshisha University graciously allowed me to present some of my early research on Japan's meat history. I am especially grateful to Professors Gavin Campbell and Ikeda Keiko for their collegiality and support. After I began to draft parts of this book, Professor Donald Stull kindly read and commented on sections focused on Japan's meat history. In 2010, I was fortunate to be able to present the Richard Geiger History Lecture at St. Ambrose University in their series "The Ubiquity of Work." In that lecture, I discussed an early version of chapter 4 on the state and meat. I am honored that the lecture committee members, Professors Ryan Dye, Daniel La Corte, Keri Manning, Sandy McKinley, and Larry Skillin, chose me for this opportunity, especially since I had a chance to see and visit with my undergraduate professor and longtime friend, Richard Geiger. Sadly, Richard has since passed away, but his interest in my work always meant a great deal to me.

My colleagues in the department of history at Western Michigan University have also provided me with help and support. I would especially like to acknowledge Professors Buddy Gray and Yoshida Takashi for their interest in discussing aspects of German and Japanese history as they relate to meat and food issues. Western Michigan University's Office of the Provost awarded me much-appreciated sabbatical support in the academic year 2008–2009 and in the fall semester of 2015. I am also grateful for additional financial support from WMU's Faculty Research and Creative Activities Award in the 2008–2009 academic year. WMU's Resource Sharing Center unfailingly filled every one of my dozens of requests for all manner of books related to meat. As with my previous book on meat and the Midwest, I could have not completed this project without the help and support of the fine staff in the Resource Sharing Center.

At the University of Iowa Press, four anonymous reviewers provided incisive comments that helped me to refine my narrative's focus. I am indebted to Catherine Cocks, the press's editorial director during the period that I worked on this manuscript, for her guidance and support. Susan Hill Newton then oversaw the project from manuscript to book. I am also grateful to Rebecca Marsh for her careful copyediting and thoughtful suggestions for improvements.

I am fortunate to have good friends who have listened to me talk at length about this project. Thanks especially to Gordon Andrews, James Cousins, Cheryl Lyon-Jenness, and Curt Miner. I also thank Nancy Spicketts for her love and support.

My three children, John, James, and Katherine, have been my greatest source of pride and joy. We shared many adventures together in Japan in 2008 and 2009. All three grew from adolescence to adulthood during the years I worked on this book, and each knows the many highs and lows that we faced as a family during that time. John has been a vegan for several years now, and he and I have had many discussions about issues related to meat and veganism that have helped me think about several parts of this book. I am proud of his ethical commitments and grateful to all my children for their love and support. I dedicate this book to them.

# Introduction

HUMANITY FACES MONUMENTAL AND MULTIFACETED environmental and public health challenges in the twenty-first century. As population continues to climb, more attention than ever is focused on the human activities that have contributed to this growth as well as its attendant dangers. Although often neglected in the array of hazards that the media publicize, many experts argue that industrial agriculture poses one of the greatest threats to our ability to sustain a clean and healthy planet. Meat production and consumption are the most significant of all the agricultural and related economic and dietary pressures that impact the human condition and the planet. Increasing numbers of books, articles, and movies highlight this "ecological hoofprint." Yet rarely do these exposés place the history of meat production and consumption in its full context. Meat eating has both benefited humanity as well as emerged as one of the most significant challenges to people's health and the planet's well-being. Understanding this spectrum of benefits and costs is central to understanding meat's history.[1]

Human beings have always eaten meat. Yet for nearly all of humanity's existence, meat was not a central component of people's diets. Starting in the latter half of the nineteenth century, meat assumed increasingly greater importance in people's diets in Western Europe and in several countries where Western Europeans had settled. Although, depending on social class, significant variations existed in the amounts of meat consumed, all people in the West, rich and poor, ate more meat in the late nineteenth century than ever before. As Western expansion around the globe accelerated during the nineteenth and twentieth centuries, promotion of meat production and consumption accompanied Western political, economic, and cultural influences. During the twentieth century, especially after World War II, virtually all cultures in the world, with the notable exceptions of those in sub-Saharan Africa and South Asia, adopted meat as an increasingly important component of their diets. Even today in India, where the average annual per capita consumption of meat is about eight pounds, many observers see a rapid transition to meat eating underway in the next generation. The dietary transition in East Asia from the mid- to late-twentieth century is especially remarkable. Japan and China transformed themselves from virtual vegetarian nations to major meat consumers within the span of two generations.[2]

This book is an attempt not only to chart the rise of meat around the world over the past two centuries and the consequences of this development but also to explain how and why this development occurred. Today, just as billions of people in Asia are gaining the means to eat more meat more often, people in Western Europe and North America are reconsidering their meat-heavy diets for reasons of personal and environmental health. If these trends continue, then a surprising reversal of the usual patterns of the past few hundred years may occur. To understand what is happening now, it is necessary to examine how the economic, political, and cultural changes of the nineteenth to mid-twentieth centuries contributed to the older dietary pattern and how developments since World War II transformed it.

One of the most common assumptions about meat eating is that through the ages all people wanted to pursue and increase it. Much of the literature on the topic of meat eating assumes that once incomes were sufficient and prices of meat products became reasonable, then consumption increased regardless of the nation or culture that people lived in. More specifically, the industrialization, urbanization, and increasing affluence that developed rapidly in the West in the nineteenth century and then spread throughout much of the rest of the world in the twentieth century typically suffice to explain the rapid rise of meat production and consumption, though some note that the relationship between affluence and meat consumption no longer exists in contemporary developed countries. This economic rationale has become axiomatic among not only economists but many other scholars. Merrill Kelley Bennett's study of wheat in national diets, published in 1941, led to formulation of Bennett's Law: an inverse relationship exists between the percentage of total calories derived from cereals and other staple foods and per capita income.[3]

Significantly, food regimes theory, developed especially by Harriet Friedmann and Philip McMichael, implicitly relies on Bennett's Law in explaining the ramifications of increased demand for meat from the late nineteenth century through the present. Friedmann and McMichael's theory contends that capitalism's development since the late nineteenth century depended in part on international connections in food production and consumption. The theory posits three stages of development. During the first food regime, Western European nations imported grain and meat from their settler states between 1870 and 1914. The first food regime disintegrated after World War I when nations erected tariff barriers. The second food regime emerged in the early post–World War II era with the development of trade guided by the General Agreement on Tariffs and Trade that was strongly influenced by US agricultural exports and a new monetary order centered on the dollar. Beginning in the 1970s but most clearly evident with the emergence of the World Trade Organization in 1995, transnational corporations replaced states as the promoters of agricultural trade policies.[4]

To be sure, this economic theory is powerful, but it does not fully explain how or why increasing affluence led to greater consumption of meat. Especially in the non-Western world, and even among segments of the West's population, meat was often not esteemed or desired. The history of meat's growing popularity in the modern era demonstrates the importance of efforts to promote it, particularly through science and by the state, as an invaluable contributor to human health. The discovery of protein and the emphasis on animal protein's central contribution to human health gave scientific support to what had long been popular belief. State support for controlling livestock diseases, especially zoonotic diseases between livestock and people, was essential for alleviating concerns about the dangers of eating meat and for facilitating trade in livestock and meat. States also increasingly regulated meat production to build public support for meat consumption. Industries of various kinds made animal products—food, fiber, and other substances—integral to various aspects of human endeavors.

Indeed, during its heyday immediately before and following World War II, meat eating was seen as a defining characteristic of civilization. As trade in feed grains from North America increasingly allowed nations in Europe and Asia to build up their livestock and meat industries and as the meat that had formerly been traded primarily among Western nations started to reach new markets outside the West, meat increasingly became a part of global consumer diets. Yet even as global meat eating became a reality in the late twentieth century, affluent Western consumers grew more alarmed by the dangers to human and animal health, as well as the larger environment, posed by increasing meat production and consumption. Understanding both the rise and the increasing ambivalence about meat in modern history requires examination of political, scientific, and cultural factors well beyond economic issues related to industrialization, urbanization, and consumer affluence.

Before attempting to introduce and contextualize the history of modern meat production and consumption, disclaimers are necessary. The first is about the literature consulted for this project. Since the nineteenth century the literary output about meat and its attendant inputs and consequences, just in English, is staggering. This book attempts to survey and glean meaning from important examples of this literature published primarily in English but in no way makes any claims about being all-encompassing. Additionally, this book is focused on the history of red meat—beef, pork, mutton, and lamb—and poultry. As defined by the US Department of Agriculture (USDA), red meat refers to cuts from animals that have more myoglobin than the flesh of chicken or fish. While fish has been an important element of the animal protein consumed in many societies for centuries, it was typically not as highly valued. The French lawyer and epicurean Jean Anthelme Brillat-Savarin wrote in *The Physiology of Taste* in 1825 that "[i]t is agreed that fish is much less nourishing than red meat, whether it is because it contains no osmazome [i.e., flavor extractives] or because being much lighter in weight for the same bulk, it has less substance." He also conveyed typical European bias by adding that fish-eating peoples are "less brave than others who live on meat."[5] In both Western and Eastern religious traditions, fish was generally not considered to be the same as meat. In addition, this book does not explain the importance of dairy products in modern history. Certainly around the globe—and like meat, starting in the West—the growing popularity of dairy products is an important part of a full understanding of the tremendous increase in protein consumption in human diets over the past century or so. But dairy, like fish, has almost always occupied a less prestigious role in this development. This means that India, the most important dairy-producing nation in the world today, is not examined in this book.[6]

Although the total literature focusing on aspects of meat's history is vast, the literature that has attempted to provide overarching assessments of meat's global production and consumption in the modern era is limited. Two books stand out in this regard. Richard Perren's *Taste, Trade and Technology: The Development of the International Meat Industry since 1840*, published in 2006, is especially significant. Perren's book assesses global meat production, consumption, and trade, and the role of states in these efforts. Perren's book, as well as his many other essays on meat, particularly on the history of meat production, consumption, and trade in the United Kingdom, have been significant resources for this project. Vaclav Smil's *Should We Eat Meat? Evolution and Consequences of Modern Carnivory*, published in 2013, is an important attempt to assess the long-term effects of meat eating. Starting with the formative role of meat in human evolution, particularly for early humans' brain development, Smil's book illuminates the rise of meat around the globe in the modern era, how meat production has impacted the environment and human health, and the ways in which meat might continue to play a role in twenty-first century diets. This book has benefited much from Smil's *Should We Eat Meat* and from many of his other publications dealing with meat issues.[7]

*Meat Makes People Powerful: A Global History of the Modern Era* explains the rise of meat production and consumption and the various effects of these developments both historically and thematically and is divided into three sections that span the nineteenth through the twenty-first centuries. This organization provides a trajectory of meat eating's importance that places its multifaceted effects in their historical context. The book's order and progression take their cue from Sidney W. Mintz's *Sweetness and Power: The Place of Sugar in Modern History*, published in 1985, which in many respects is still one of the most influential books written about a particular food's multifaceted economic, social, and cultural impacts in the modern world.[8]

Part I is focused on the foundational divisions between the meat-eating and limited meat-eating parts of the world during the nineteenth century. Chapter 1 examines the foundations of Western meat eating, urbanization's effects on meat demand, creation of an international meat trade among the West and its settler nations (which was part of the first food regime), Western settler nations' exports to Europe, and the relative magnitude of Western nations' meat consumption before 1900. Chapter 2 focuses on the limited meat cultures of East Asia, specifically Japan and, to a lesser extent, China, the two most important limited meat-eating nations in the modern era that have transformed into significant meat-eating societies over the last half of the twentieth century. The chapter explains the reasons meat eating was limited and then charts how meat became gradually more available to consumers by the early twentieth century.

Part II builds on part I's mostly economic focus by examining how science and state policies supported the growth of meat production and consumption from the mid-nineteenth through the mid-twentieth century. While part II's focus is primarily on Western developments, the section also examines Japan and China and the consequences of these Western promotions on Eastern meat cultures. Chapter 3 explains how the discovery of protein and a growing consensus about its health benefits were rapidly marketed to lower income people in the West and elsewhere through meat extracts and canned-meat products. Meat also became a central part of the eugenics debates during the early twentieth century. In Japan, eugenics arguments helped to promote meat eating as essential to national development. Chapter 4 surveys the major ways that national governments in the West and Japan promoted meat consumption. One of these was increasing regulation of meat trade and oversight of meat safety. Governments around the world intervened to ensure that livestock diseases were contained, that meat could be traded around the globe, and that these products would be safe for consumers to eat. Although governmental policies varied in the degree to which they promoted either public or private production of meat, the basic desire was similar: reassure the public that meat was safe and clean and thus suitable for consumption. Chapter 5 explains how beef, long considered the most prestigious meat, became the focus of consumer demand through the mid-twentieth century. At the same time, meat industries increasingly marketed inedible animal by-products as a means not only of improving the profitability of meat production but of ensuring public support for the value of animal industries. By the World War II era, a general convergence in meat-eating patterns and consumption of by-products had developed, which in turn would facilitate even greater production, consumption, and trade after the war.

Part III examines how, after World War II, meat became both a global food and, increasingly, a target of consumer concern. Chapter 6 explains the fundamental ways in which the meat trade that emerged after the war intertwined with the second food regime, which was dominated by US agricultural and economic policies. In particular, the US feed grain trade allowed nations in Europe and Asia to develop their own livestock and meat industries. Western meat trade networks were also expanded as a part of the new era of freer global trade. As meat became a global food, most nations' consumers adopted poultry over beef as their primary meat choice. This poultry revolution has continued to shape meat-eating patterns into the twenty-first century. Trade and poultry issues are explored in more depth in terms of the rise of meat eating in Japan and China.

The remaining chapters of part III explore how global meat eating has created growing ambivalence and alarm. Chapter 7 focuses on the evolution of perceptions about the meat industry's workers as participants in a killing industry. While both the East and West had cultural traditions that tended to ostracize meat workers, state intervention during much of the twentieth century helped to dissipate many of these prejudices. However, the erosion of state protections for meat workers during the late twentieth century and the emergence of a global labor market combined with the animal rights movement have given rise to renewed prejudices about meat workers. Even more than workers, the two areas of greatest consumer concern about global meat production and consumption are its effects on the environment and human health. Chapter 8 examines two of the most important global environmental concerns: water pollution and air pollution. Chapter 9 summarizes the public health concerns that stem from late twentieth century research on meat's nutrition and dangers from meat microbes. Increasingly, governments are encouraging consumers to limit their meat consumption. Indeed, China and Japan have emphasized the need for their citizens to look back to their traditional diets as a basis for health. The evidence conveyed in chapters 8 and 9 illustrate how the role of science and the state has shifted in recent years from firm support of meat's value to increasing caution.

Chapter 10 concludes the book by assessing current projected meat consumption trends in light of the concept of sustainability. According to projections through 2050, continued increases in global meat production and consumption are not sustainable, especially given the likely environmental and human health effects. Increasing educational emphasis on sustainable livestock and meat production may make possible meat substitutes more acceptable and help to stabilize meat production and consumption at more reasonable levels.

# PART I

# Foundations

## Meat Eating in the Nineteenth Century

# 1

## Nineteenth-Century Meat Cultures

THE WORLD'S MAJOR MEAT CULTURES DEVELOPED initially in Europe and parts of the world colonized by Europeans. Britain, Germany, and several urban centers in Western Europe had emerged as prominent meat producers and consumers by the mid-1800s. During the same era, European settlers established frontier livestock cultures in North and South America and Oceania, particularly in the United States, Canada, Argentina, Uruguay, Australia, and New Zealand. In each of these countries, the immigrant populations and their descendants consumed significant amounts of meat and their growing livestock industries expanded to supply Europeans with increasing amounts of meat. These developments established the Western patterns of meat production and consumption that would then expand to new parts of the world during the twentieth century.

A combination of agricultural, dietary, economic, and political factors propelled the development of Western meat cultures. Livestock herding and improvement traditions in which cattle, hogs, and sheep were bred partly for consumption dated back hundreds of years. While quantities were limited for most of the population, meat was considered essential to good health and was also an indicator of social standing. As urbanization expanded in Western Europe during the nineteenth century, changing government policies supported new slaughtering and processing facilities, transportation systems, and marketing mechanisms that resulted in meat markets in major cities such as London, Paris, and Berlin, which supplied their expanding populations with ever increasing amounts of meat. Rising numbers of urban Europeans expected to be able to purchase regular supplies of meat. The significant upsurge in meat demand began in Britain and the German states' urban centers then spread throughout the countries' general populations during the last quarter of the 1800s. To meet this mounting consumer demand, partly because domestic sources declined in parts of Europe, an international meat trade among Western nations developed with major suppliers in the United States, Canada, Australia, New Zealand, Argentina, and Uruguay, all of which then became leading meat cultures in their own right. By the start of the twentieth century, the factors that had contributed to the Western meat cultures would start to make inroads in many other parts of the world.[1]

## Western Meat Cultures' Foundations

At the beginning of the nineteenth century, meat was a part of all Europeans' diets. Indeed, meat in many forms had been an important part of Europeans' diets going back hundreds of years. Yet throughout Europe before 1800 meat was only a regular and substantial part of the diets of the most affluent. As Jeffrey M. Pilcher notes, "Before the nineteenth century, only the elite could enjoy white bread and meat on a daily basis."[2] Vincent J. Knapp affirms the same idea: before 1800, meat consumption in Europe was "mainly a matter of privilege."[3] This pattern had been in existence since the Middle Ages. In the 1300s, for instance, English peasants primarily subsisted on cereals, though evidence suggests that some of the more affluent peasants increased their meat intake during the fourteenth and fifteenth centuries. Beef consumption also increased starting during the mid-1300s in many European towns. Whereas bacon had been the only type of meat consumed in the 1200s, by the 1300s and 1400s, many had access to fresh beef and mutton. Some historians argue that meat's availability surged somewhat in the aftermath of the Black Death. The relative paucity of meat then returned, however, by the later sixteenth century. Yet even in the fairly narrow window of time in which more meat was available, nearly 80 percent of the average European's diet in the fourteenth century consisted of grains in the form of bread, oatmeal flour, and ale. This pattern held true for the next four centuries. During the early modern period, Europeans primarily subsisted on bread or other foods made from grain, including beer. They ate only small quantities of vegetables, fruit, and meat. Across the globe before industrialization, animal sources provided less than 15 percent of all dietary protein, and saturated animal fats provided around 10 percent of all food energy. In 1800, for example, meat provided less than 3 percent of all food energy for the French population.[4]

In Europe, and especially in the European settled rural and frontier areas, seasonal availability limited fresh meat consumption and encouraged meat preservation. This fact continued to shape meat preferences into the twentieth century. Preserved beef was most popular in Argentina and Uruguay, while preserved pork was most popular in the North American colonies and early United States, though when available North and South Americans still preferred fresh beef. For instance, affluent colonial New Englanders ate about as much beef as pork. In North America, where hogs were allowed to forage more frequently than cattle, hogs were easier to rear and required less feed. Preserving the meat through salting and smoking helped to stretch the limited amounts of meat.[5] As Sarah F. McMahon explains, "early New Englanders did not expect a year's provisioning from the annual slaughter in the early winter. Many seventeenth and early eighteenth-century New Englanders assumed that their barrels of salt pork and beef would last only through the late spring."[6] Throughout South America, cattle foraged for their food. In Brazil's Rio Grande do Sul and Argentina's and Uruguay's Rio de la Plata regions, salted or jerked beef, called *charque* in the former and *tasajo* in the latter, was a cheap form of preserved meat exported to the masses in Cuba and Brazil.[7]

Although meat consumption was limited for most Europeans until the late nineteenth century, important traditions of raising livestock for food were significant in providing a foundation for meat cultures in Europe, the Americas, and Oceania. Two of the world's foremost cattle herding traditions developed in the British highlands and the southwestern Iberian Peninsula and were then transported to the New World along with European settlement. Both of these herding traditions contributed to the North and South American livestock frontier's development. Indeed, British entrepreneurs were significant investors in cattle ranches in both the United States and Canada. The Iberian longhorn that was the basis for cattle ranching in the lowlands around Seville became the basis for the cattle stocks that populated Latin America.

Spanish animal herding traditions, particularly of cattle and sheep, influenced nearly all of Latin America's livestock cultures beginning in the early sixteenth century, first in the Caribbean Islands, then Mexico, before affecting regions further south. During the seventeenth century, the Spanish turned cattle, horses, and sheep loose in the regions around Brazil's Rio Grande do Sul as well as Argentina's and Uruguay's Rio de la Plata. Although not considered high quality animals, the cattle in these areas thrived and were utilized for a variety of purposes by the romanticized and hardscrabble Argentine and Uruguayan gaucho and Brazilian vaquero. During the first half of the nineteenth century, wealthy Argentine landowners, known as caudillos, also turned to cattle ownership. Beginning in the mid-nineteenth century, these owners imported English cattle, primarily Shorthorn bulls, to improve the quality of their native cattle. Soon after, British Herefords, Aberdeen Angus, and Galloways would also help to improve the cattle breeding stock in North and South America. Spanish Merinos as well as French Rambouillet and German Negretti breeds of sheep were introduced in Argentina, Uruguay, Australia, and New Zealand during the mid- to late-nineteenth century to improve the existing stock. European swine had traditionally been reared as mast forest scavengers. But starting in the 1600s in Britain, due to deforestation and growing urban populations, they were increasingly interbred with Chinese hogs. By the end of the eighteenth century, these improved hogs resulted in the development of Yorkshire, Berkshire, Hampshire, and Suffolk breeds that would also become significant in improving the swine populations of North America.[8]

While Europeans and European settlers in the New World enjoyed the taste of meat, they were also convinced that meat was essential to good health.[9] Evidence for these beliefs is particularly clear in commentaries where Europeans describe their encounters with New World indigenous peoples and their diets. For instance, Spanish settlers in Latin America were appalled by native people's consumption of insects, reptiles, and toads. For the early modern Spanish, "[b]read, wine, oil and meat provided proper nourishment for the body, and were the basis of health."[10] In England, Europe's greatest meat-eating nation, mutton had been more widely consumed among all classes of people through the early modern era, but after the mid-1600s beef consumption rose steadily. Increasingly during the eighteenth century, roast beef on Sundays became a common practice among the wealthy.[11] This habit's importance may be best enshrined in Henry Fielding's patriotic song, "The Roast Beef of Old England" (1731): "When mighty Roast Beef was the Englishman's food, It ennobled our Hearts and enriched our Blood."[12] A Swedish observer, Peter Kalm, noted in 1748, "I do not believe that any Englishman who is his own master has ever eaten a dinner without meat."[13] Roast beef and beef in general—typically roasted, boiled, enclosed in pastries, and made into other dishes—symbolized affluence, strength, health, and confidence, all elements that British and British settlers around the globe by the late nineteenth and early twentieth centuries increasingly identified with as well. During the same period in France, overindulgence in meat was often associated with carnality and sexual excess, but meat was nevertheless also believed to be essential to good health.[14]

English people's desire for beef grew rapidly over the span of the nineteenth century. This meat preference then accompanied English settlers to their new homes in North America, Australia, and New Zealand. Swine numbers in England declined from the late seventeenth through the eighteenth centuries. Industrialization and urbanization spurred a population explosion, which in turn fueled vastly higher demands for meat; between 1600 and 1800, England's population doubled from four to eight million and then nearly quadrupled between 1800 and 1900 from eight to thirty million. Friedrich Engels notes in his history of the English working class in 1844 that, when employed, the better-paid workers ate some meat every day, while the lower-paid workers ate meat two or three times a week and sometimes only on Sundays. Only the poorest ate no meat at all. Although during the modern era, Germans, as was true of the British, generally prized beef roasts above all other meat dishes, in 1870 the most widely used of all meats throughout the various parts of what became Germany were pork dishes, both fresh and cured. Pork was the most popular meat in North American diets through the nineteenth century. Yet all but the very poorest Americans ate meat at every meal.[15]

## Urbanization and Growing Meat Demand

Several developments contributed to the increasing meat production and consumption that accompanied urban population growth in Europe during the nineteenth century. These factors included rising incomes, social conditions associated with urban lifestyles, expansion of free trade, improved transportation as well as new food processing and manufacturing industries, and increased imports. New regulatory policies regarding meat production and consumption were also significant. Throughout continental Europe until the end of the eighteenth century, guild butchering traditions had dampened both meat production and consumption. France abolished guild control over the butcher trade in 1791, which then ushered in a period of relatively unfettered meat production until Napoleon I established municipal slaughterhouses outside of Paris, partly to address citizens' concerns about pollution. By the mid-nineteenth century, various degrees of government support for meat production throughout Europe were commonplace and essential to providing ever greater quantities of meat that was safe to eat.[16]

Paris and London were the two largest centers of meat consumption in nineteenth-century Europe. In the case of Paris, as well as many other cities across Europe, removal of traditional regulations helped to increase urban residents' meat consumption. For instance, in Paris in 1858, the abolition of registration and price controls led to a three-fold increase in the number of beef butcher shops from 501 in 1856 to 1,805 in 1875. Pork butchers, who had not been similarly restricted, also increased in line with the city's population expansion during the same period, from 422 to 654. Increasing availability of meat then helped to solidify urban Parisian and Londoner's beliefs that meat was crucial to proper diet and good living even though the quality and quantity of meat eaten among the various classes in both cities varied considerably. Yet nearly any urban resident ate considerably more meat than someone living in the countryside of either France or England. Beef became the most important type of meat consumed in both cities.[17]

Construction of modern public slaughterhouses in European cities also contributed to meat consumption's increase over the span of the nineteenth century. Following Paris's lead, where five public slaughterhouses were established on the city's periphery in 1810, Rouen's first modern public slaughterhouse opened in 1830, followed by Brussels in 1840, Marseilles in 1848, Vienna in 1851, Lyon in 1858, Frankfurt in 1861, Milan in 1863, Munich in 1865, Zurich in 1868, Hamburg in 1872, and Berlin in 1881. Continental Europe's greatest central public slaughterhouse and meat market complex, La Villette, opened in Paris in 1867. Although construction of public slaughterhouses was also tied to urban consumer's concerns about government protection for public safety and health—a topic explored in chapter 4—these facilities contributed to the declining cost of safe meat. The situation in Berlin illustrates this combination quite well: after 1883, all meat slaughtered in Berlin or imported from elsewhere had to be inspected for public safety and health at the central public slaughterhouse. As the city's population grew from 824,484 in 1871 to 1,864,779 in 1900, the number of animals slaughtered nearly doubled. By 1890, Germany had ninety-four public slaughterhouses. In Britain, the public slaughterhouse movement occurred later and was more contentious than on the European continent. Nevertheless, private slaughterhouses and butchering expanded dramatically over the nineteenth century.[18]

Marketing and supplying meat to North Americans depended primarily on private enterprise. In New York City, cattle were driven or transported by ferry from outside the city to local stockyards on the Lower East Side, where butchers often slaughtered and marketed beef in public markets. Starting in the 1820s, merchant-wholesalers based along the Ohio and Mississippi River towns of the Midwest provided increasing amounts of meat to the rest of the nation. During the antebellum period, Cincinnati became the nation's chief meat-producing city. Among the nation's leading Midwestern packing towns, Cincinnati in the mid-1840s slaughtered one-quarter of the hogs that were sold at market. Through the early post–Civil War years, US meat production remained largely tied to fall and winter slaughtering and focused on mostly salted and cured products, especially pork. Before the development of mechanical refrigeration in the late nineteenth century, pork was more popular to preserve than beef. The South American preference for jerked beef did not gain traction in North America. Indeed, the American meat industry continued to focus on pork products like ham, bacon, and sausage in the 1920s and 1930s partly because most Americans did not have home refrigerators that would allow them to purchase fresh beef.[19]

Nevertheless, as railroad expansion shifted the American meatpacking industry to Midwestern cities with better rail connections, access to fresh beef became increasingly available to urban consumers. With the opening of the Union Stockyards and Transit Company in south Chicago in February 1865, Chicago became the earliest and most important terminal and transfer market for meat products. Soon after, several other major stockyards opened in major Midwest cities, including Kansas City, Omaha, St. Louis, Milwaukee, Indianapolis, and St. Joseph (Missouri). Before the end of the nineteenth century, large stockyards also opened in Wichita, South St. Paul, and Sioux City. In addition to rail connections, growth of meat production also hinged on developments in mechanical refrigeration, of which the most important was the refrigerated railcar. Beginning in 1878, Gustavus Swift developed the first successful fleet of refrigerated railcars specifically designed for what is called the dressed beef trade. That is, quarter carcasses of beef were shipped by rail from the slaughterhouses to branch-house distribution centers. Cold-storage facilities were also developed along the key distribution points, which helped to create nationwide meat markets. These facilities were also essential to the development of European exports. The firms known as the Big Five—Swift, Armour, Morris, Cudahy, and Schwarzschild and Sulzberger (which became Wilson in 1916)—had 571 branch houses by 1900. Most of these firms supplemented branch-house distribution with peddler-car service that allowed salesmen to sell meat to consumers in isolated locations throughout the country. By World War I, the Big Five companies' branch house and refrigerated car routes reached twenty-five thousand cities and towns in the United States.[20]

Meat, leather, and wool production were central to economic development in Argentina and Uruguay during the eighteenth and nineteenth centuries. Through the 1870s, Argentina's sheep and cattle herds produced the great bulk of the country's wealth. Millions of cattle each year were converted into *tasajo*, dried and salted hides, and tallow, 95 percent of which went for export: about half the tallow went to England and much of the *tasajo* was sold for slave food in Brazil and Cuba. The opening of the Société de Fray Bentos Giebert & Cie, at Fray Bentos in Uruguay in 1863, which then became the London-based Liebig's Extract of Meat Company in 1865, signaled both a continuation of older uses of the region's relatively poor quality cattle as well as a new step in the direction of increasing internationalization of the region's beef industry. Liebig's Extract was a spreadable condensed beef product used as a cheap source of protein that fueled new demand for animal protein around the globe, as explained further in chapter 3. By 1875, the Fray Bentos plant produced over 1.1 million pounds of extract, which then tripled in volume by 1891. Although limited in amounts because of their small populations, beef not sold abroad was consumed domestically and made Argentina and Uruguay into two of the world's great meat-eating nations.[21]

## Creation of an International Meat Trade

Until the mid-nineteenth century, Britain's domestic livestock sources satisfied urban consumers' meat demands. In Leeds, for instance, where the population more than doubled between 1801 and 1831, cattle marketed there came primarily from the north, especially Scotland and Northumberland County, while sheep came from the east, especially the Yorkshire Wolds. At London's Smithfield Market, opened around AD 950, the volume of livestock handled from interior locations was much greater. As per capita meat consumption in London expanded from 70 pounds per capita in 1750 to between 122 and 153 pounds one century later, Smithfield grew beyond its capacity to handle the numbers of cattle and sheep shipped there. Even with rail shipments growing rapidly after the 1830s, the closest railhead was at Euston Station, about two miles from Smithfield. As a consequence, livestock needed to be herded through busy urban areas from the railway station to market. Before the slaughtering portion of its facilities closed in 1855, Smithfield Market handled a peak of 329,000 cattle in 1853 and 1.66 million sheep in 1851.[22]

Primarily due to livestock diseases, domestic beef production in Britain stagnated between 1840 and 1870. Pleuropneumonia was discovered in the nation's cattle in 1842 and was endemic thereafter. Foot-and-mouth disease also spread during the same period. Millions of cattle were lost due to disease during midcentury. Because of urban consumers' growing demands for more meat at reasonable prices, increasing numbers of livestock, both alive and dead, were acquired from foreign markets by in the mid-1800s. Before 1870, Britain looked especially to Ireland as a source of meat supplies. In 1842, restrictions on imports of livestock were lifted in Britain, and Irish cattle, sheep, and, especially, swine were routed to London's meat market. As Irish grains, cattle, pigs, and dairy products increasingly went to England from the 1840s on—exacerbating the effects of the Great Famine—Ireland's agricultural economy became devoted to export trade, particularly livestock production. British meat markets also pursued livestock sources in the United States and continental Europe. In 1870, Great Britain imported 10 percent of its total meat supply, mostly from Ireland and continental Europe, with a small proportion from the United States. Although less important in supplying meat overall, the United States had supplanted Ireland beginning in the 1830s as England's supplier of preserved pork and bacon.[23]

After 1870, market factors, disease concerns, and, especially, technological developments shaped Britain's increasing reliance on beef supplies from overseas. Largely because of British demand for meat and other agricultural products, the first food regime developed, tying suppliers in North and South America as well as Australia and New Zealand with the growing European market demand for meat and for grain staples such as wheat. After 1870, market factors, disease concerns, and, especially, technological developments shaped Britain's increasing reliance on beef supplies from North and South America as well as Australia and New Zealand. Largely because of British demand for meat and other agricultural products, the first food regime developed. This regime tied suppliers in North and South America as well as Australia and New Zealand with the growing European market demand for meat as well as grain staples such as wheat. British beef supplies came from both live cattle and chilled or frozen beef carcasses. Canned-meat products also played a part in the international meat trade. British markets received their first live cattle shipments from the United States in 1868. For the rest of the nineteenth century, live cattle slaughtered in Britain brought higher market prices, primarily because consumers believed fresh products tasted better and were safer. Consequently North American live cattle predominated over chilled beef through the end of the nineteenth century. A new outbreak of European pleuropneumonia in 1877 spurred a jump in American and Canadian live cattle imports. When pleuropneumonia was detected in American cattle at the Liverpool market in January 1879, Britain's Privy Council ruled on March 3, 1879, that, under the provisions of the Contagious Diseases (Animals) Act, live cattle from the United States had to be slaughtered at port facilities within ten days. Canada was not placed under the same restrictions until 1892, which helped to triple the amount of its live cattle exports between 1880 and 1891. During this period, animal products constituted one-third of total Canadian exports. Ireland also continued to provide a considerable amount of Britain's live cattle. During the 1890s, Irish contributions made up close to 20 percent of Britain's total meat supply, which was double the proportion from a generation earlier. Australia, Argentina, Uruguay, and New Zealand contributed much of Britain's remaining imports at century's end. By 1900, Britain's domestic meat supplies contributed only about forty-five pounds of meat per capita, which was down from sixty pounds in the 1860s.[24]

In contrast to Britain, Germany increased its meat supplies in the late nineteenth century primarily through aggressive domestic agricultural expansion supported by Chancellor Otto von Bismarck's protectionist economic policies. In 1879, Germany relied on American pork products for only 10 percent of its domestic supplies. Nevertheless, American pork exports had helped to bolster supplies for the growing demand for meat in Germany. Discovery of American hams infected with trichinae by a University of Vienna professor in 1879 set off the German-American Pork War of 1880 to 1891. As explained in chapter 4, whether disease or economic concerns were the primary catalyst in the curtailment of American and other imported meat, the pork war was part of a larger protectionist trend that supported both German aristocratic landowners as well as peasant farmers, the primary pig producers in the country, from the 1880s through the World War I era. After 1900, meat imports made up less than 5 percent of the German meat supply compared to more than 50 percent in the case of Britain. Because of huge domestic increases in production, however, pork prices remained steady. While between 1800 and 1913 the number of cattle in Germany doubled from 10 million to 20 million, the number of hogs increased sixfold from 3.8 million to 25 million. Nevertheless, Germans were beginning to diversify their tastes to other meats before the war and were also willing to pay more for meat at every price level.[25]

## European Settler Nations' Meat Exports

New technologies, most specifically shipboard mechanical refrigeration, made the export of chilled and frozen meat products from North and South America as well as Oceania possible. After the first chilled beef arrived from New York City in 1875, US chilled beef became an increasingly important part of British meat supplies. This was due in part to the fact that Britain's laboring poor were buying more meat and could generally only afford cheaper chilled products. Pork continued to be slightly more important overall than beef among American meat exports to Britain. By the start of the twentieth century, the United States supplied nearly 30 percent of Britain's total meat supplies. Half of this total consisted of pork, and beef constituted almost all the other half. Chilled mutton made up only a very small percentage of the total. American products constituted very high percentages of certain categories of British meat imports, specifically salted beef, bacon, and hams.[26]

To an even greater degree than the US meat industry, Canadian meat production was oriented primarily toward the British market by the end of the 1800s. The William Davies Company in Toronto was Canada's largest meatpacker in the nineteenth century. It produced cured pork products, particularly those known as Wiltshire sides, which were cut and smoked for British consumer preferences. Ian MacLachlan points out that the foundation of the Canadian meat-packing industry was bacon. Nevertheless, some canned meats, such as corned beef, were also produced in the 1890s for the British market. Through the nineteenth century, the majority of Canadian cattle were raised for live export rather than local consumption.[27]

Starting in the 1860s, several of the Australian colonies, including Victoria, New South Wales, South Australia, and Queensland, began exporting significant amounts of canned meat to Great Britain, initially because of greater British demand due to ongoing livestock disease problems and new developments in canning technologies. Also important was the British Admiralty's adoption of canned meat as a general ration in 1846. Unlike the case in Latin America, Australia's meat industry was largely homegrown. While private initiatives were central to the industry's development, the colonial government provided bonuses and subsidies, extension and advisory services, and help with establishment of cold storage and refrigerated transportation. Until the 1880s, New Zealand was the major external market for Australian meat and manufactured goods in general. However, once New Zealand became self-sufficient and a host of technical obstacles were overcome, Great Britain became Australia's major external market. Several meat-canning plants opened in the 1860s and 1870s, and the canned-meat trade with Britain reached its peak in the early 1880s.[28]

The pioneer of meat freezing for export in Australia was Thomas S. Mort, who established the first freezing works at Darling Harbor in Sydney in 1861. Attempts to prepare meat for export failed initially, but the first successful shipment of mutton and beef sent from Australia finally occurred in 1880, when the SS *Strathleven* landed the first consignment of frozen meat in London. Technological developments in both the loading and receiving ports were required in addition to shipboard advances. After 1882, chilled and frozen meat products from Australia started to regularly arrive in Britain. In 1884, 138 tons of Australian beef and 3,176 tons of mutton reached Britain. Frozen meat did not really take off in sales until the 1890s. By 1893, 10,549 tons of beef and 14,358 tons of mutton from Australia reached British markets.[29]

New Zealand became Britain's leading supplier of frozen lamb and mutton starting during the 1880s. Between 1881 and 1883 alone, eleven meat-exporting firms emerged. New Zealand's meat exports then increased by 3,000 percent between 1880 and 1890. The country would maintain this position in the British meat supply chain into the 1920s. In the nineteenth and early twentieth centuries, British direct investment in New Zealand's meat sector was considerably less than in either Australia or Argentina. Most of the New Zealand meat companies were domestically owned and generally of a smaller scale than those in Australia or Argentina. After 1900, however, British firms owned many of New Zealand's freezing plants for meat exports.[30]

After the quality of Argentina's and Uruguay's cattle and sheep stock improved in the 1880s and 1890s, both countries became significant contributors to the meat trade with Britain. Argentina's fresh beef exports lagged considerably behind competition from the United States, Australia, and New Zealand in 1895–99; but its mutton exports surpassed Australia's and were not far behind New Zealand's. Supported by Argentinian government incentives, four *frigoríficos*—refrigerated packing plants—were founded before 1902. British investment was late but was also crucial to this development. By 1908, British firms and individuals owned nearly 30 percent of the land in Santa Fe province in Argentina, and British investment was a central part of frozen and refrigerated meat trade investment there. Still, compared to American chilled meat, Argentina's frozen meat was at a competitive disadvantage in the European market in this period. Argentina did not develop a successful chilled beef sector until 1901, and the country's *frigoríficos* were not financially successful until the start of the new century. Importantly, when British ports closed on April 30, 1900, to Argentine live cattle because of foot-and-mouth disease, this event stimulated the *frigoríficos* to devote more resources to chilled beef. Fresh beef production finally surpassed traditional jerked beef production in 1904. Much of Argentina's frozen beef and mutton was marketed to lower income consumers in the Midlands region of England. During the late 1800s, Uruguay exported almost no frozen mutton and only began to export significant amounts of frozen beef during the early twentieth century. However, Argentina's and Uruguay's exports to Britain rapidly expanded before World War I. Except for the 1917–18 period, when the United States supplied 80 percent of Britain's meat and animal fats, Argentina became Britain's major foreign supplier of beef and mutton before the war and then through the Great Depression era.[31]

## Western Meat Consumption before 1900

European settlers in the New World not only brought their cultural proclivities to consume meat with them, they benefited from several factors that made many of their countries into the greatest meat consumers of the period. In the case of Australia, New Zealand, Argentina, and Uruguay, rapid urbanization combined with ready access to their frontier regions' growing livestock industries were especially important. At the end of the nineteenth century, Australia was the world's leading meat-eating country. Average estimated meat consumption ranged from 246 pounds per capita in 1890 to 208 pounds in 1895 to 262 pounds in 1900. These amounts were nearly double those of the United States during the same period. Because Australia's population was relatively urbanized from an early point—nearly 30 percent of its people lived in a handful of coastal cities by 1850; and 38 percent of the populations of Victoria and New South Wales lived in Melbourne and Victoria by 1891—canned meat also helped to make meat available at low cost to greater numbers of people. Indeed, one of the problems with investment in meat-canning plants in Australia was that investors were hesitant to put money in operations aimed primarily at working-class people, both in Australia and Britain. Nevertheless, it seems likely that canned meat's availability was one of the factors in establishing the very high levels of meat consumption evident in Australia in the nineteenth century.[32]

Although it resembled Australia in its cultural composition and rapid urbanization, New Zealand's overall population, and resulting domestic meat demand, was smaller. While Philip S. E. Hereford estimated that meat consumption before 1900 was about 130 pounds per capita, Yves Guyot's estimate for the decade of the 1890s was 212 pounds per capita. Although still less than Australia's levels during the same period, meat consumption was likely higher than that for both the British and American population and possibly even second highest in the world. Somewhat lower demand, though, meant that the country did not develop meat canning to the degree that Australia did in order to supply its burgeoning domestic demand.[33]

Likely third highest in average world meat consumption by 1900, Americans may have actually eaten more meat in the antebellum era than they did at the turn of the twentieth century. As early as the 1830s, US meat consumption, with pork significantly outpacing beef, was estimated to be 178 pounds per capita annually. Immigrants from Britain and Germany, Europe's two greatest meat-eating cultures, were especially prominent in the growing US population. Although pork continued to be the dominant meat consumed by Americans at the turn of the twentieth century, greater availability of beef year-round through the

Big Five's extensive distribution system transformed Americans' meat-eating habits. Fresh beef then became the American meat of choice. As part of a more general transformation of American diets during the late nineteenth century, the average American by 1900 was consuming more milk, fruits, and vegetables as well as beef. In the decade of 1891–1900, Americans on average consumed about 150 pounds of meat and about 170 carcass-weight pounds of meat during the subsequent decade.[34]

Although more difficult to determine on a national level compared to European and North American countries, Argentina was possibly fourth among the world's highest average meat consumers. The growing number of lower and moderate income urban consumers in Rio de Janeiro purchased significant amounts of Argentina's *tasajo*, though about 35 to 45 percent of the country's product continued to go to Cuba in the early 1890s. Beef jerky competed with salt cod as the leading plantation food. Although the freeing of Brazilian slaves in 1888 curtailed that market for beef jerky, sales of *tasajo* did not peak until 1895. Most of this demand came from the growing urban population in Argentina and Uruguay. Urban consumption of fresh beef in Argentina and Uruguay started to climb only in the early twentieth century.[35]

Britain's meat consumption rose rapidly from the mid-nineteenth to the early twentieth centuries, and it was probably the greatest meat-consuming country in Europe as the twentieth century began. In 1861, national per capita meat consumption was 59.5 pounds. By the 1880s, it was about 110 pounds, and by the first decade of the new century, it had reached 130 pounds. Social historian John Burnett estimated that between 1903 and 1913, it averaged 135 pounds. Although before the 1850s, about three quarters of Britain's meat consumption consisted of beef, between 1875 and 1900, beef made up half of the meat eaten, followed by pork with about 30 percent, and mutton about 20 percent. Social class divisions nevertheless remained quite significant in British meat consumption. In 1900, wealthy Britons ate on average around 300 pounds of meat per year, including typically the great bulk of the fresh meat available, the middle class ate an average of 162 pounds, while the poor ate about 100 pounds, made up mostly of products derived from frozen and chilled meat imports as well as preserved pork, such as bacon. J. L. Van Zanden's study of European agricultural laborers from 1870 to 1914 found that in 1870, British agricultural laborers had consumed quite a bit more meat than any other Europeans. However, by 1910, this group of British poor had slipped behind their French, German, and Swiss counterparts in average annual meat consumption. Each of the latter ate just over one hundred pounds per year while British agricultural laborers consumed slightly less than one hundred pounds.[36] As Reay Tannahill put it, "British factory laborers existed on a diet of bread and tea, with a piece of cheap meat or bacon on Sundays."[37] Yet even Britain's poor ate more meat than almost any other nation's people in the world at that time.

As in Britain before 1800, meat was typically scarce in all but the wealthiest Germans' diets. In Prussia in 1816, the estimated annual meat consumption was about thirty pounds per person.[38] Lizzie Collingham notes, "The standard meal of an eighteenth-century German rural labourer was 'gruel and mush,' a soupy combination of grains and lentils."[39] Social status both determined and reflected how people consumed meat. In contrast to the largely beef and mutton meat culture of Great Britain, Germans were traditionally more focused on pork; and particularly after the mid-nineteenth century, greater pork production allowed Germans to expand their meat consumption. In 1855, average meat consumption was about 43 pounds per capita. At the turn of the twentieth century, Germany was the world's sixth-largest meat-eating country, with an average of one hundred pounds consumed per capita; and the large numbers of peasants involved in the nation's agriculture during the late nineteenth century specialized in raising animals, especially hogs. The importance of pork consumption among Germans is also evident in the fact that in 1894 it was the most important component of meat consumption in all regions of the country with the exception of Prussia and the Rhineland, where beef was most important. In 1907, pork made up 60 percent of Germans' meat consumption compared to 49 percent of American, 28 percent of British, and 17 percent of French consumption. By 1900, average meat consumption in Germany was about 88 pounds per person before it increased to nearly 100 pounds by 1914.[40]

Several common elements characterized the West's significant meat cultures during the nineteenth century. While it is difficult to sort out which of these factors was most significant, all of them were interrelated to some degree. In each of the cases explained in this chapter, a long-standing emphasis on livestock rearing is evident. In the European settler states, these traditions were transplanted early on and cultivated in concert with their European counterparts. In each, meat eating was clearly associated with good health and appropriate cultural behavior and aspirations. At the same time, as people in these various places became more affluent, they also demanded beef over all other meat choices. As rural to urban movements transformed Western Europe over the eighteenth and nineteenth centuries, urban residents cultivated an ever greater desire for meat as part of their standard of living. Urbanization was also a key factor in creating access to meat in the European settler nations. Rising incomes made increased meat consumption possible, and eating meat was also a conspicuous indicator of growing affluence. Important technical changes made this dietary transition possible, especially burgeoning rail transportation and mechanical refrigeration, both on ships and in manufacturing and commercial locations. Meat was a central part of the trade associated with the first food regime. Particularly in the case of the British people, their desire for ever-greater quantities of meat required acquisition of new sources of meat production in North and South America as well as Oceania. These developing societies provisioned meat not only for British and other European markets but also in huge amounts for their own peoples. These patterns stood out in stark contrast to the world's limited meat cultures, particularly in East Asia.

# 2

# Nineteenth-Century Limited Meat Cultures

MEAT'S EMERGENCE AS AN IMPORTANT GLOBAL DIEtary element is linked to many developments in world history, including urbanization, affluence, and international trade. Widespread meat eating has in turn had multiple impacts on global cultures, some of which are obvious, such as on health and the environment, and some less obvious, such as reinforcement of social status prejudices. For most non-Western societies, eating meat was limited well into the twentieth century. When meat consumption became a more frequent dietary component after the mid-twentieth century, it involved choices that signaled the acceptance, though not wholesale adoption, of elements of Western cultures. Perhaps the starkest transitions from limited to considerable meat eating over the course of the nineteenth and twentieth centuries occurred in East Asia, particularly in China and Japan. Both these nations struggled with the incorporation of meat into their cultures and did so in significant fashion only during the last half of the twentieth century.

A complicated combination of religious, ethical, political, economic, geographic, and agricultural factors contributed to these nations' status as limited meat-eating countries. Religious and ethical traditions, particularly stemming from Buddhist beliefs, suppressed meat consumption throughout East Asia. Poverty was just as important in limiting meat eating. Meat was simply too expensive for the East Asian masses. Before Buddhism became established in China, neither of the country's traditional ethical belief systems, Confucianism and Taoism, banned meat consumption. Indeed, meat eating was expected during important social events and feasts. On other occasions, sacrificial rituals discouraged meat consumption. In general, only the wealthy were exempted from minimal meat eating. Buddhism's permeation throughout Chinese culture after the Han dynasty helped to further minimize meat consumption, particularly beef, because of the Buddhist focus on animal compassion. Then both Confucianism and Taoism gradually adopted more of a Buddhist approach in their teachings about meat. Through the mid-twentieth century, farm animals in China were valuable for their traction power and generally much too important to eat. Use of draft animals for agricultural traction power did not peak in China until the mid-twentieth century.

Likewise, Japan's transition from a society with virtually no meat consumption to one that consumed some meat was fitful and protracted. The early Chinese commentaries from the third century AD on Japanese society already noted the lack of meat eating there. Both Buddhism and Shinto, the native religious belief system, helped to suppress meat consumption. By the late seventh century AD, a ban on beef, horse, and chicken consumption was put into place. More so than in the case of China, the state's role in propelling Japan's adoption of a wider array of meats is particularly important. Beginning during the Meiji era (1868–1912), when the new Japanese government and many sectors of Japanese society embraced Western technologies, ideas, and ways of thinking, meat was also seen by many Japanese to be a necessary component of the nation's growth and development as a modern society. During the succeeding Taishō (1912–1926) and Shōwa (1926–1989) eras, government leaders continued to pursue policies that emphasized the importance of meat eating and supported the necessary infrastructure that would encourage it. Although Japan did not emerge as a significant meat-eating society until later in the Shōwa era, developments that began during the Meiji era, particularly in terms of government-supported infrastructure and policies in the areas of livestock development, consumption promotion, and slaughterhouse regulation, laid the groundwork for accelerating adoption of meat in Japanese diets following World War II.

## Foundations of Limited Meat Cultures in East Asia

Early East Asian peoples domesticated and ate a wide variety of animals. The earliest evidence of pig husbandry in China dates to about nine thousand years ago in the middle Yellow River Valley of north China. Millet agriculture developed independently there as the subsistence base. People ate both fish and pig meat for protein. The early Chinese then domesticated dogs and sheep. Later chickens and goats were likewise domesticated. By the time of the Shang dynasty, nearly three thousand years ago, all these animals, as well as cattle, horses, and water buffalo, were important. Yet few Chinese ate much animal meat. Traditional Chinese religious belief systems, specifically Taoism and Confucianism, did not prohibit meat eating per se. Taoist dietary practices focused on prohibition of foods that might interfere with bodily harmony. Unlike Taoism, while Confucianism emphasized compassion for animal life, it mandated no general condemnation of animal slaughter for human consumption. Both Taoism and Confucianism seem to have adopted a greater sensitivity about meat eating later on, though primarily in reaction to Buddhist teachings. Before the end of the Han dynasty, when Buddhism had gained a significant foothold in China, the most important religious influence in suppressing meat consumption was sacrificial food culture. While the rich and nobility were allowed the privilege of eating meat obtained from hunting, the lower social orders were deprived of meat, except during special occasions when chicken might be consumed.[1] Roel Sterckx emphasizes the central importance of meat avoidance in early Chinese history: "Meat abstention reflected ritual purity and moral sincerity and, as such, the intake of meat was to be minimized in preparation for many rituals, during fasts, and—most importantly—during the traditional mourning period."[2] Thus sacrificial food culture and poverty contributed to Chinese people's overall lack of meat consumption in its early history.

Buddhism's contributions to vegetarianism in China became significant beginning in the fifth century AD as a result of two key factors: monastic regulations and Mahāyāna scriptures. The monastic regulations said that monks could eat meat only if they did not kill the animals themselves or eat meat killed specifically for them. While the scriptures maintained that killing animals was wrong, the emphasis was primarily on the consequences of killing for the individuals who consumed the meat rather than on the animals' suffering. Through the onset of the Tang dynasty in the seventh century AD, even after the Emperor Gaozu ordered that animals could not be killed for three months each year and for ten specified days, meat eating was still expected during certain social rituals. Concerns about meat eating, as had been true during earlier periods, were tied to its association with the wealthy and their luxurious and immoral habits of indulgence.[3]

Traditional Japanese religious and cultural practices likewise suppressed meat consumption. A Chinese chronicle, *Wei zhi* (*The History of the Kingdom of Wei*), written between AD 220 and 265 and compiled about 297—which was approximately four hundred years before the official ban on meat eating—noted Japanese concerns about eating animals. Influenced by Buddhist ethical concerns about animals, as well as Shinto ideas about ritual purity and cleanliness, Emperor Tenmu announced a ban on the eating of beef, horses, and chickens in AD 675. The ban did not prohibit eating deer or wild boar, which were the most important animals in the Japanese diet during this period. Similar edicts continued to be pronounced between AD 676 and 781. While wild boar and venison continued to be eaten, especially during the celebrations surrounding the new year, the ban limited overall meat consumption. The ban was meant in part to help preserve the limited stock of draft animals used for traction power, a concern shared by the Chinese people as well.[4]

During the Song dynasty from the tenth to the thirteenth centuries, China's agricultural and food systems began to assume their modern contours. Rice replaced millet as the chief grain, and the regional cuisines started to take shape. Buddhism's contribution to both Taoism's and Confucianism's increasing attention to vegetarianism also emerged. Over the course of the ninth to the fourteenth centuries, Buddhism developed a beef taboo that would last into the twentieth century. Due primarily to traditions of respect for animal life and other religious and socioeconomic changes during this period, the beef taboo also became part of Taoist texts from the Song dynasty onward. The beef taboo held true even among the elite.[5] Yü (Princess) Der Ling noted the beef avoidance among the imperial court's members in the Forbidden City during the early twentieth century: "Beef was a thing that was tabooed within the precincts of the Palace, as it was considered a great sin to kill and eat animals that were used as beasts of burden. The food consisted mostly of pork, mutton and game, fowls and vegetables."[6] The Chinese beef taboo did not directly impact Japan's similar taboo, but they were parallel and resembled each other.

Although China during the Yüan dynasty was ruled by the Mongol Empire and some food items took on more of a Central Asian flavor, little changed in terms of limited meat consumption through the Yüan and Ming dynasties. Mongols consumed mare's milk and butterfat and mutton, but these items did not gain traction in the general Chinese diet, though, to be sure, tremendous dietary variety resulted from the geographical and agricultural differences in the country. Wheat in the North and rice in the South became the staple foods for China's growing population. Indeed, China's population grew by 50 percent from the end of the fourteenth through the mid-seventeenth centuries, and the Chinese were likely the best fed people in the world based on diets of cereal grains and various beans. By the end of the Ming and throughout the Qing period, New World foods also permeated the Chinese diet. Sweet potatoes, for instance, became a staple food, especially among the poor in the coastal southeast areas of the country by the early 1700s. As it had been for nearly two millennia, meat was consumed in sizable amounts only during banquets and rituals. For the typical Chinese, meats, particularly pork and chicken, were eaten primarily as flavorings or as part of sauces with rice and vegetable dishes.[7]

While Japanese government policies and cultural beliefs upheld the ban on eating cattle, horses, and chickens, wild game and fish could be consumed. Contrary to strict Buddhist tenets, and often despite typical Buddhist practices in China and elsewhere in Asia, Japanese did not consider fish as meat in terms of dietary restrictions.[8] Basil Hall Chamberlain, one of the prominent nineteenth-century English commentators on Japan, believed that allowing fish eating was "a concession to human frailty."[9] Paradoxically, too, some segments of the Japanese population consumed some beef. For example, the *eta* (handlers of dead animals) and *hinin* (social outcastes plus impoverished peasants or townspeople), who collectively came to be known as *Burakumin* or *Buraku*, especially after the Meiji era, both produced and consumed beef and horsemeat. Social outcastes are central to any discussion of Japan's meat history, both past and present. Although speculation about the *Buraku*'s origins has evoked a tremendous amount of disagreement among scholars, it seems fair to say that those who handled dead cattle and horses existed as a recognized social group long before the Kamakura era (1185–1333), probably dating back to the seventh century. They were both further stigmatized and classified as part of Japan's lowest social order with the control ordinance of 1591. Scholars seem to agree about the worsening of the various social outcastes' status during the Tokugawa era. Some point out that certain tasks performed by *eta*, such as stripping leather from horses and cattle, were done without pay during the Tokugawa era. Indeed, part of the enduring social stigma attached to the perceived social outcastes was the fact that people were appalled that they not only made leather but that they ate the meat from such animals.[10]

Nevertheless, in both China and Japan, despite the array of social and cultural restrictions that limited meat consumption, long-held beliefs that meat played a role in maintaining good health persisted. In this respect, East Asian traditions shared similarities with European traditions. In Japan, though official statements and public pronouncements supported a ban on eating certain domestic livestock through the end of the Tokugawa era, some meat continued to be eaten by a fairly wide range of Japanese society and not simply among the *eta* and *hinin*. Eating wild game, for instance, was said to be *kusurigui* (eating medicine). Consumption of meat was often, as one historian describes, "cloaked under euphemisms."[11] Indeed, Shimizu Akira goes so far to say that by the late 1700s "[t]he use of euphemisms seems to have encouraged more and more people to visit the beast market [in the Kōjimachi district of Edo (Tokyo)] to experience meat dishes because such euphemisms exempted them from admitting that they actually ate meat."[12] In late Tokugawa Edo, wild game was sold in shops called *momonjiya*, which advertised with a maple tree sign for deer or a peony tree for wild boar. The most famous of these shops was a butcher shop called *Kōshūya* in the Kōjimachi district of Edo, which made virtually all types of wild game meat available by the late 1700s to mid-1800s. Such shops were also referred to as *okuyama* (deep in the mountains) in reference to the wild game's origins. The meat sold, particularly wild boar, was called *yamakujira* (mountain whale) or *yamakurage* (mountain jelly). Japanese peasants particularly favored wild boar meat, partly because its high fat content was thought to impart medicinal properties. Throughout premodern Japanese society, the sick were encouraged to eat meat due to popular beliefs that it contained nutrition that provided physical strength. Beef by-products were commonly sold as medicine. Before being banished during the early Tokugawa era, Christians, prominent in places like Nagasaki, which also had a long history of various meat dishes, ate meat. The well-known Japanese Christian daimyo, Takayama Ukon, was said to be a beef eater. Of course, the Tokugawa rulers could also use Christians' beef eating as a sign of their barbarity and the need to be rid of them. Some evidence suggests that some of the Tokugawa elites ate beef as well. According to one Japanese historian, a Russian delegation that visited Nagasaki in 1853 offered ham and mutton to some local samurai who eagerly ate the meat and asked for more.[13]

Before the Meiji era, animal slaughter occurred in isolated (mostly *Buraku*) locations and generally not in large numbers at openly acknowledged sites. Japanese historians claim that the only site of open cattle and horse slaughter during the Tokugawa era was the trading outpost island of Dejima in Nagasaki, which was associated with the Dutch in the city. Toward the end of the Tokugawa era, though, animal slaughter occurred in several urban areas. The establishment of foreign settlements beginning in the 1860s was probably the most important motivation for greater meat production and consumption. During this period, it seems likely that the organizers of many of the meat and dairy production sites were businessmen, not *eta*. For instance, in the Edo region, a foreign settlement was established in Yokohama in November 1864. Maeda Tomekichi established a dairy in Yokohama to provide milk to foreigners. When the Shogunate and the delegations from the United States, Britain, France, and the Netherlands signed a commercial treaty, it included an agreement to allow cattle slaughter. Japan's first public slaughterhouse opened in April 1865, shortly after the memorandum was signed. Initially, cattle destined for the Yokohama plant were sent from the Kansai region, especially the Ōmi area (today's Shiga Prefecture), via the Kobe port by Japanese mail ship. This seems to be the origin of Kobe beef. The beef industry in Edo was made possible in part by the same trade networks in place by the early eighteenth century that allowed merchants in Ōsaka to ship agricultural and manufacturing products efficiently to Edo's consumers. By 1868, seven firms in Yokohama imported and sold meat from animals shipped from Kobe. In 1869, the Ministry of Finance established *Gyuba Kaisha* (Cattle and Horse Company) for producing and selling beef and dairy products. Although foreigners were major customers, increasing numbers of urban Japanese were eating meat, though often for medicinal reasons.[14]

## Increasing Meat Availability

Over the course of six months in 1878, during the midst of Japan's intensive modernization period, Isabella Bird, an English pastor's single, middle-aged daughter, who had already traveled in Australia, Hawaii, and the American West, visited Japan. In the book that recounted her travels there, Bird expressed repeated misgivings about Japanese food. Because her views so blatantly revealed her prejudices about the superiority of a Western diet, her observations about Japanese food, particularly because she primarily traveled in rural areas of the country, point out the contrasts between food in Japan and much of the West at this time. The "Food Question," as she described it, comes up very early and is repeated frequently in her account. "The fact is that except at a few hotels in popular resorts which are got up for foreigners, bread, butter, milk, meat, poultry, coffee, wine, and beer, are unattainable, that fresh fish is rare, and that unless one can live on rice, tea, and eggs, with the addition now and then of some tasteless fresh vegetables, food must be taken, as the fishy and vegetable abominations known as 'Japanese food' can only be swallowed and digested by a few, and that after long practice."[15]

In a telling follow-up to these remarks, Bird added, "After several months of traveling in some of the roughest parts of the interior, I should advise a person in average health—and none other should travel in Japan—not to encumber himself with tinned meats, soups, claret, or any eatables or drinkables except Liebig's extract of meat." Bird felt that the majority of the population had little access to quality food. "The diet of the poorer classes is meagre and innutritious, revolting in appearance and taste, and the quantities of sauces and pickles with which they render it palatable are very injurious to the digestive organs." She lamented the lack of meat in Japanese cuisine. For example, Bird noted that during her visit to the isolated northern Honshu village of Shinjō in Yamagata Prefecture that a Japanese companion had finally secured a chicken for her to eat. She complained about suffering from days of a meatless diet and a "lack of such food as one can eat when one finishes a hard day's journey without appetite, in an exhausting atmosphere." Although favorably impressed with the physical condition of the Ainu people she met in Ezo (Hokkaidō), in contrast to her opinion of the Japanese, she noted that their diet was also mostly "abominable things." However, the Ainu often ate wild game, especially venison and bear meat.[16]

Although Bird did not reveal any awareness of this fact in her travel account, increasing numbers of Japanese, mostly in urban locations, were eating meat during the 1870s. But it was still an unusual experience in the overall scheme of Japanese food culture. In addition to religious and ethical obstacles, another significant roadblock to Japan's development as a meat-eating society was the nature of the country's deeply rooted agricultural emphases. At the beginning of the Meiji era, Japan's primary agricultural focus was rice production. Government policies attempted to modify this emphasis somewhat. Nevertheless, early Meiji initiatives stressed production of basic food staples, especially rice and, to a lesser degree, developing marketable surpluses. Relatively little effort had been directed toward livestock development before the Meiji era. The Tokugawa government encouraged daimyo and other landlords to raise cattle and horses for military transportation purposes.[17]

Beginning in the 1870s, however, the Meiji government started to implement policies that gradually shifted Japan's agriculture toward more diversified farming, including livestock production. Increasing livestock production, combined with the emperor's decision to lift the ban on meat eating, was part of what Japanese historians refer to as *bunmei kaika*, "Civilization and Enlightenment" policies that promoted Westernization. In terms of promoting livestock farming, the government supported importation of livestock animals from North America, Europe, and East Asia. In 1881, the Ministry of Agriculture and Commerce was established, which in 1925 was reorganized into the Ministry of Agriculture and Forestry. To support animal breeding, particularly the crossbreeding of foreign imports with native animals, such as *wagyū* cattle, it launched animal disease control measures and disseminated information about stock-raising techniques. The government published a variety of literature supporting livestock farming, such as *Bokuchiku Shinkō Ikensho* (*Recommendations for the Development of Animal Husbandry*) in 1881, and established supportive organizations, such as *Chikusan Shijunkai* (advisory council for the livestock industry) in 1884. Government policies encouraged livestock farming in regions of the country where it had not traditionally been practiced. For instance, cattle breeding had taken place in the Tokugawa era in Tottori, Shimane, Hiroshima, the Chūgoku mountains, Okayama, and Hyōgo. The government encouraged cattle production in Kantō, Tōhoku, and Hokkaidō. For instance, between 1883 and 1910, cattle numbers jumped from virtually zero to over 20,000 in Hokkaidō and from about 10,000 to over twenty thousand in Chiba, just north of Tokyo. The Meiji government also established cattle breeding stations in the early twentieth century in Nanatsukahara, Tsukisappu, and Ōita that remained active until 1916. When demand for chicken meat and eggs rose following the Sino-Japanese War (1894–95), tariffs were imposed on these products in 1902 to protect domestic production. An agricultural experiment station for poultry production was established in Sakura, Chiba Prefecture. The Meiji government's efforts in terms of research and extension were also encouraged at the prefectural government level; and, beginning in 1899, subsidies were provided for prefectural experiment stations. In addition, the Meiji government removed the feudal restraints on farmers' decision making that had been part of Tokugawa era agriculture. Individual farmers' innovations were instrumental, for instance, in developing improved cattle breeds and eliminating the need for government-directed cattle breeding stations. Between 1880 and 1900, livestock production grew at a rate of 6.8 percent per year. Nevertheless, by the end of the 1920s, the proportion of farms with animals present either for labor power or meat production still remained quite limited.[18]

Meiji agricultural diversification efforts emphasized not only importation of Western livestock but also recruitment of Western livestock-rearing expertise. For instance, William Smith Clark, a professor at Massachusetts Agricultural College who had befriended several important Japanese students there, was hired in 1876 to help establish a new agricultural college in Hokkaidō called Sapporo Agricultural College. One of Clark's principal projects at the college was building a model farm that included, among other elements, cattle and horses that could help to develop animal husbandry skills among the students. Working with Edwin C. Dun, a cattle farmer from Ohio, Clark encouraged the college to import Ayrshire dairy cattle from California and develop native *Nanbu* cattle as draft animals for the college farm.[19]

The Meiji government was equally concerned with encouraging meat consumption. Arguably the first step in this process was the issuing of the Emancipation Edict in August 1871. This edict officially abolished the status system in place since the Tokugawa era, but, as *Buraku* historian Suginohara Juichi said, "[D]iscrimination against *Burakumin* did not disappear," in part because the government provided no administrative measures that enforced social equality.[20] Such measures would not be enacted until after World War II. Nevertheless, the edict officially severed the country's old association between *Buraku* and livestock meat consumption. The emperor then gave meat eating a further boost on January 24, 1872, when he lifted the ban on its consumption and then ate meat publicly. Although the emperor had apparently first tasted meat in late 1871, after January 1872, the emperor was said to eat beef and mutton on a regular basis. Later in the spring of 1872, the emperor issued a proclamation that allowed Buddhist priests to eat meat, much to the dismay of many Japanese.[21]

Many Japanese were not immediately receptive to eating meat, and some harbored suspicions about doing so for decades after this event. After public outrage about the opening of a new beef shop in Fukui Prefecture in October 1872, which in 1881 became the modern prefecture of Shizuoka, the prefectural government admonished the people's prejudices. Japanese meat historian Hirai Kiyotaka describes the event:

> A beef shop was opened in Tsuruga, but people gossiped about many problems with the shop. The prefectural government issued a document that said beef supplements people's health and that it contained nutritive medicine. There are people who adhere to the old traditions and avoid eating beef. There are people who still believe that beef is unclean. These behaviors obstruct civilization. The gossip is untrue. If there is someone who still believes this gossip, please come to the prefectural office. We will explain the truth.[22]

Some of the resistance to beef consumption reflected social class biases. In addition to its long-standing association with the *Buraku*, Meiji era meat shops were often located in unsavory parts of cities. Fukuzawa Yukichi, Japan's most ardent popularizer of meat during this period, founder of Keiō University, and author of the first Japanese book about Western food, wrote during this period that beef shops in Tokyo were frequented by hoodlums whose bodies were covered in tattoos. Beef shops were often thought to be situated too close to prostitute quarters. Nevertheless, meat shops became increasingly common in Tokyo—Edo's new name beginning in 1868—and other cities throughout Japan during the Meiji era. By 1875, Tokyo had almost as many beef shops (seventy) as bird-meat and fish shops (seventy-seven).[23]

Meat advocates like Fukuzawa underscored the government's promotion of meat eating as an indicator of civilized behavior. Fukuzawa's influence was evident in some of the signs on butcher shops in Tokyo that read "meat for your health" or "government-approved meat." For those like Fukuzawa, meat symbolized civilization, sophistication, and social prestige. Opponents of meat consumption were seen as standing in the way of progress and promoting backward superstitions. Even a well-known skeptic about Westernization's benefits, Kanagaki Robun, the pen name of Nozaki Bunzō, author of the famous parody called *Aguranabe* (*The Beefeater*), gave meat eating a backhanded promotion when he said, "[Y]ou won't get civilized if you don't eat meat!"[24] Another meat promoter, Andō Tarō, claimed in his defense of meat eating, published in 1897, that some Japanese opposed meat consumption for three main reasons: religious concerns, willful ignorance, and lingering beliefs that meat should be eaten only by those who are sick. Andō believed that many of the religious concerns reflected long-standing fears of Christianity. He said that those who felt meat was only for the sick needed to understand that its nourishing qualities would benefit all.[25]

The Meiji government also encouraged the military to provide beef for its soldiers. In addition to playing a crucial role in stimulating increased need for horses, cattle, and sheep for transport and clothing purposes, the military included beef in its soldiers' diets. Canned beef was introduced in the army in 1877 at the time of the Satsuma Rebellion, and the army later opened its own canning factories. The navy made meat part of its rations with passage of the Food Supply Act in 1884. *Yamatoni*, beef in soy sauce and sugar, apparently became a mainstay for soldiers during the late nineteenth century. Canned beef products were seen as an indicator of both progress and modernity. The Japanese navy made beef a central part of its sailors' rations in the 1890s despite objections from some officers and enlisted men. In 1898, Japan submitted bids to American meatpackers for supplying canned and salted meats for use in the army and navy. American meatpacker Swift and Company, for instance, opened distribution branches in both Tokyo and Ōsaka. Military service became one of the main vectors in Japanese society at this time for creating support for beef consumption.[26]

While government support was crucial for the expansion of meat production and consumption, private initiatives were also significant in increasing both livestock supplies and slaughter facilities. In 1869, Kanagawa Prefecture entrepreneur Kakuta Yonejirō established a pork farmer association, *Kyōkyūsha*, which encouraged establishment and expansion of pig farming, as well as development of swine stocks from overseas. He had established a pig farm in Yokohama with seventy to eighty boars. Kakuta is reported to have promoted his efforts with the slogan: "Rice is a food for adults and emperors. Poor people should eat meat." Not surprisingly, then, he also is reported to have encouraged *Buraku* to raise pigs. As contrasted with the movement of cattle from the Kansai region to Tokyo, hogs went the opposite direction from Kanagawa to Kansai via the Tōkaidō road. In 1875, Tokyo had four legal sites for hog slaughter: Yanagihara in Honjō, Amari-machi in Ōkubo, Asakusa Shintani-machi, and Shirogane. Some hog slaughter occurred in open areas and in residential areas.[27]

Cattle rearing in the Ōmi area had already been underway during the late Tokugawa era, and Ōmi remained an important center for cattle during the Meiji era. Takenaka Hisaji was a rice wholesaler who became the leading beef wholesaler in Tokyo by promoting Ōmi beef. The lore surrounding him emphasized how he walked cattle from Ōmi to Tokyo along the Tōkaidō road and often fought off bandits trying to steal his cattle. *Nanbu* beef from Iwate Prefecture in northern Honshu also became popular during this period. *Nanbu* cattle had been bred as transport animals during the Tokugawa era and then were interbred with American Shorthorn cattle that were imported into the area after 1871. Another entrepreneur, Horikoshi Seijirō, was credited with opening the Tokyo beef market to *Nanbu* beef in either 1872 or 1873.[28]

Sales of beef in Tokyo depended on the availability of natural ice. Nakagawa Kahei established the first ice company for this purpose in Yokohama in 1872. His company depended on ice transported during winter from Mt. Fuji and then on ice from the mountains in Shinshū in modern-day Nagano Prefecture. Both connections were short-lived. Nakagawa then established a factory in Hakodate in Hokkaidō, which made ice balls that were transported via ship to Tokyo. The Tokyo Ice Production Company initiated the first mechanical ice production in Tokyo in October 1883, but ice of any sort remained expensive throughout the Meiji era.[29]

Horsemeat, also banned by Emperor Tenmu in 675, was reintroduced not long after beef and pork gained popularity. Sales occurred as early as 1874–75, and Noguchi Yoshitaka received a license to sell horsemeat in Ueno in Tokyo in 1877. Although consumption of horsemeat did not carry the same stigma as it did in many Western countries, shop owners nevertheless distinguished sales of beef from horsemeat with different colored flags. The distinction was apparently meant to heighten consumer interest in buying beef. Red or orange flags were hung outside shops that sold beef while black flags signaled shops that sold horsemeat. Government authorities also fined shop owners that sold horsemeat advertised as beef.[30]

Until the early twentieth century, animal slaughter occurred primarily in privately owned facilities. Nevertheless, government authorities exercised considerable regulatory control over private facilities starting in 1871. This regulatory power became especially important during the early twentieth century. The earliest laws focused on inspection of beef slaughterhouses. Laws also licensed beef shop owners. The history of Tokyo's meat production during the Meiji era demonstrates steady if erratic availability. Between 1869 and 1906, when the Slaughterhouse Law took effect, slaughterhouses were repeatedly opened and closed, often by the government for reasons pertaining to public complaints about poor sanitation. Tokyo's Cattle and Horse Company failed seven years after its opening in 1869. Yet between 1877 and 1886, the number of cattle slaughtered in Tokyo jumped from 7,025 to 21,985. In December 1893, Iwatani Shōhei, Kimura Shōhei, Suzuki Genzō, and Takenaka Hisaji consolidated five private Tokyo slaughterhouses into the Japan Domestic Animal Market Company to gain greater control over meat production. Yet neither this private sector development nor government efforts to better manage production and quality stopped rapid private sector development in the nineteenth century; by 1894, a total of 903 slaughterhouses existed throughout Japan. Tokyo had four major slaughterhouses in 1897: Shirogane, Minowa, Tanakamachi, and Mikawajima. Later, at the end of the Meiji era in 1912, Tokyo's five major slaughterhouses were Ōsaki, Minowa, Terashima, Hachiōji, and Fussa. Japan's two major private meat companies today, Itoham and Nippon Meat Packers, did not originate until the Shōwa Era (1926–1989)—Itoham in 1928 and Nippon Ham in 1942.[31]

Ōsaka was the other early major center of Japan's meat industry. Its early meat production history illustrates the growth and chaotic nature of the industry as well as its connection to the *Buraku* during the Meiji era. *Buraku*, concentrated in Kizu and several small villages that later became part of the modern Nishinari neighborhood of the city, had primarily slaughtered deer during the winter before the 1870s. In the early Meiji era, a larger slaughter facility, as in the case of Tokyo, was built to provide beef for foreigners who resided in Kawaguchi in 1872 and 1873; and many *Buraku* butchers were hired to provide laborers for the facility. During the decade of the 1870s, beef eating increased enough in the region to spur further meat production. By 1879, four private slaughterhouses in the Ōsaka area killed just over two thousand cattle and horses. An even larger facility, owned by the Ōsaka Slaughterhouse Company, was established in June 1887. By 1902, the plant had adopted some mechanization though refrigeration capabilities were apparently absent since all meat slaughtered in the morning had to be sold before noon that same day. Another significant private packing plant located in Tennōji neighborhood, the Naniwa Slaughterhouse Limited Partnership Company (later, Kansai Slaughterhouse Trade Company), opened in 1894 and closed in 1905.[32]

One feature of Japan's expanding meat production that had no parallels in the West was incorporation of Buddhist temples onto the grounds of packing plants for the purpose of praying for the dead animals' souls. For instance, the Japan Livestock-Raising Industry Company established one such temple at its Minowa plant in 1909, while another temple was established at the Ōsaki plant. The Minowa temple was later transferred to the public slaughterhouse at Shibaura when it opened in 1936. At such temples, which would become common at public slaughterhouses later in the twentieth century, a Buddhist memorial ceremony or requiem called a *kuyō* is performed for the souls of the dead animals. Anthropologists contend that the ceremony harks back to the reincarnation themes in early Buddhism. Their purpose is to settle the souls of the dead animals and also provide prayers for the safety and well-being of the people who slaughter them. The existence of such ceremonies also suggests the continuation of Japanese ambivalence about meat production and consumption.[33]

While the amount of meat consumed by the average Japanese in the Meiji era was still minimal, it steadily increased. In 1885, a Japanese newspaper noted that meat eating varied significantly between urban and rural locations. Comparable to the pattern in Europe during the same era, residents in major urban areas such as Ōsaka and Tokyo ate the most, while residents of smaller towns and rural areas in Hakodate and Nīgata ate very little. Most Japanese typically consumed skewered meat called *tsujiuri kushi* that was stewed in soy-based soup like *oden* or prepared in thin strips in a pan with green onion and sake called *yakinabe*. *Gyūnabe*, popular especially in the Yokohama and Tokyo area, was cut into chunks before stewed. Both were relatively cheap. For *sukiyaki*, which was prepared in a fashion similar to *yakinabe*, thin strips of beef were sautéed in a pot along with soy sauce, tofu, and a variety of vegetables and noodles. This dish became increasingly popular during this period, especially during winter. The first *sukiyaki* shop opened in Yokohama and was followed by one in Tokyo. Japanese could also purchase thin slices of horsemeat for dishes similar to *yakinabe* beginning in 1887.[34]

Increasing prosperity among Chinese during the early twentieth century encouraged more meat consumption. Between 1900 and 1924, cutting across the end of the Qing and beginning of the Republic, average wages in China quadrupled. Most types of food consumption rose dramatically as well. Pork consumption doubled. Although the increase is impressive, it is only so in the context of meat consumption that had been almost nil before this point. Between 1922 and 1925, John Lossing Buck studied 1,070 farm families who supplied over 80 percent of their food on their own farms. His study, which spanned seventeen localities across seven provinces, found that 90 percent of their food energy was derived from grains and another 9 percent from root crops, especially sweet potatoes. Animal products made up just 1 percent of Chinese food energy sources. During the same period in the United States, only 40 percent of food energy came from grains. While Chinese diets were not deficient in protein, acquiring it from grain, legumes, and vegetables meant that the quality of the protein was not especially good. According to Buck, "There is little variety in the diets, and hence a few products are consumed in large quantities."[35] Those main items consisted of rice, wheat, millet, corn, beans or peas, kaoling, and sweet potatoes. Even in the mid-1940s, grains and legumes continued to make up 88 percent of Chinese diets compared to 38 percent of those in the United States. As in the early to mid-1920s, average protein intake was not deficient but came overwhelmingly from nonmeat sources. Before 1949, when Communist China was established, meat consumption remained "strikingly low."[36]

The animals that were present on Chinese farms in the late 1920s and early 1930s provided primarily fertilizer and traction power. Although the significance of cattle for these purposes had long helped to limit beef consumption, this facet of Chinese animal husbandry attracted greater Western attention during the early twentieth century. Buck's extensive research on land utilization in China during that period found that only a small percentage of farms kept animals for meat production purposes. Just 12.8 percent of Chinese farms raised hogs for meat, followed by sheep (5.9 percent), and poultry (4.8 percent). Even pigs, which were raised for meat purposes, were just as valuable to Chinese farmers for their manure production. When consumed, Chinese ate all parts of the pig, including the offal. For instance, traditional beliefs encouraged use of hog spinal cords in preparations to treat venereal diseases and cooked pig blood was made into cakes. Some trade of Chinese hams with urban areas in the Asian Pacific, especially Manila, existed during the 1930s. Significant numbers of farms had either cattle (33.5 percent) or water buffaloes (22 percent) to use for their traction power, but these animals were too valuable to be used for meat. As Buck explains, "Farm animals in China are chiefly raised for draft purposes, but in spite of this fact the density of animal population is greater than it is in a country like the United States where the larger proportion of animals is for productive [i.e., food] purposes, such as cows raised for the production of milk."[37] Farms in the United States were already transitioning to the use of mechanical instead of draft animal power. During the mid-1930s, one estimate of Chinese meat (primarily pork) consumption was about 34 pounds per person annually.[38]

Despite increased availability, especially in larger urban areas of the country, meat eating among the Japanese also remained quite limited in an international context during the early twentieth century. Russell H. Chittenden (1856–1943), the Yale biochemist and advocate for a limited protein diet, remarked on this fact at some length in *The Nutrition of Man*, an influential study of the role of protein in human diets published in 1907. Chittenden conceded that Japanese diets typically lacked adequate protein: "the daily diet of the great bulk of the Japanese people has been characterized by a very low proteid [*sic*] standard, as contrasted with the standards and usages of the majority of European and American people. . . . The stature and general physique of the people could be advantageously improved."[39]

In both China and Japan, longstanding ethical, cultural, and economic conditions diminished meat eating among their general populations through the mid-twentieth century. Compared to the Western meat cultures where livestock were reared in large part for meat consumption, neither China nor Japan had similar traditions for such activity. By the late nineteenth century, Chinese and Japanese government efforts encouraged animal husbandry but had little effect through this period in making meat a more significant dietary component. While increasing numbers of Chinese and Japanese people associated meat eating with both good health and superior lifestyles, these ideas continued to meet resistance within the larger cultures despite government efforts to convince people otherwise. In Japan, the association of meat production with the undesirable *Buraku* status group also made people hesitant to consume it. Overcoming the various barriers to meat consumption—present in both China and Japan—was a critical factor in the successful incorporation of meat into the diets of Chinese and Japanese people after World War II. At the same time, increasing urbanization and affluence would encourage meat consumption in ways similar to how these factors had affected people in the West. Intertwined, though less obviously, were the scientific and state regulatory and trade policies that propelled meat to a greater dietary role in virtually all national diets after World War II.

# PART 2

# Promotions and Results

## Scientific and State Support for Meat before World War II

# 3

## Debating Meat's Dietary Role

As the nineteenth century progressed, Europeans and European settlers in the Americas and Oceania consumed increasing amounts of meat. Nutritional science's advance during that time provided scientific support for this diet. The developing science of nutrition reinforced and solidified popular beliefs—evident in both Western and Eastern cultures—about the strength-giving properties of meat. During the 1840s, the German chemist Justus von Liebig's investigations into the nitrogen-rich substance called protein convinced people of meat's foundational role in human diets. Perhaps even more important, his role in founding the Liebig's Extract of Meat Company helped to make inexpensive beef products available to people in many countries around the globe and create demand for beef in ways that would reinforce expanded production and consumption of beef and other meat products into the twentieth century.

Also during the late nineteenth and early twentieth centuries, a diverse array of scientists and health advocates debated the amount of protein necessary to support proper nutrition. While most of the protein supporters focused on meat's role in providing this nutrition, vegetarians argued that plant sources could provide enough protein to maintain good health. Segments of these meat and nonmeat advocates became important contributors to the eugenics movements during the early twentieth century. While many of the eugenics debates centered in the West, some of their ideas also influenced nutritional debates in other parts of the world, including Japan, where prominent medical and academic authorities used the pro-meat arguments popular among Western eugenics advocates to underscore the importance of meat in Japan's national diet. Increasing acceptance of a nutritional role for meat would prove to be one of the crucial elements in convincing East Asians to make animal protein a more prominent part of their diets. After the mid-twentieth century, new nutritional studies would temper this consensus about meat's nutritional importance and increasingly lend support for moderation, as explained in chapter 9.

## Science and the Marketing of Inexpensive Animal Protein

Justus von Liebig (1803–1873), one of the most influential chemists in world history, is particularly responsible, through both his theoretical work and practical interests, for stressing animal protein's significance for human health. Liebig's laboratory at the University of Giessen in Hesse-Darmstadt was one of the few in Europe open to students, and many of the leading researchers associated with protein chemistry worked with him. While Liebig is often credited with discovering protein and then publicizing meat as the most nutritious of foods in his 1842 publication titled *Animal Chemistry*, his investigations built on the work of Dutch chemist Gerhard Mulder, as well as the French and Belgian chemists Antoine-Laurent Lavoisier, Claude Berthollet, Antoine Fourcroy, Karel van Bochaute, François Magendie, and Jean-Baptiste Boussingault. Liebig, Mulder, and others asserted that the nitrogenous molecule that plants and animals shared should be called the proteine radical, or protein. In Liebig's words, "[v]egetables produce in their organism the blood of all animals, for the carnivora, in consuming the blood and flesh of the graminivora, consume, strictly speaking, only the vegetable principles which have served for the nutrition of the latter."[1] Translated, Liebig emphasized the protein food chain whereby plants nourish herbivorous animals, which then feed carnivorous animals, including humans. Although during the twentieth century more refined scientific findings replaced most of Liebig's specific arguments in *Animal Chemistry*, the book highlighted chemical processes of animal digestion, respiration, assimilation, and degradation in a way that captured the attention of chemists, physiologists, medical doctors, and sanitarians. Moreover, his work on the chemistry of meat caused food and cookbook writers to pay attention to science in a way that had not occurred before.[2]

Arguably just as important as Liebig's scientific work was its resulting promotion of meat teas and extracts that helped to make animal protein available to increasing numbers of people around the globe. Like others in the mid-nineteenth century, Liebig believed that meat juice was highly nutritious. Before Liebig developed his beef extract, beef tea—a cold effusion of meat—was used for restorative purposes. For instance, Liebig thought that meat teas and extract could replace soup in hospitals. Before the 1860s, however, Liebig's beef tea was not widely available. Only a few London pharmacies made it available, and typically only the wealthy purchased it as a substitute for mothers' breast milk or as a tonic for nursing mothers.[3]

Availability of meat teas and extracts improved during the 1860s. In 1862, Georg Christian Giebert, a German railway engineer employed in the Rio Grande do Sul region of Brazil bordering Uruguay and Argentina, met with Liebig in Munich to receive technical instructions on preparing beef extract. Consisting of reduced beef stock and a 4 percent solution of salt, beef extract was a thick, syrupy, dark brown spreadable form of condensed beef with a powerful aroma and a texture not unlike molasses. It was sold in glass jars and later in cans. Giebert explained that Uruguay's territory near the Rio de la Plata and Uruguay River was a growing cattle production area where often only the hides were used. Carcasses were either left to rot or processed in *saladero* plants into *tasajo*. Later that year Liebig enthusiastically supported Giebert's first venture: setting up a small factory next to a *saladero* in Fray Bentos, Uruguay. After receiving additional financial support, Giebert purchased the Fray Bentos *saladero* and founded the Société de Fray Bentos Giebert et Cie in 1863. Two years later, the company was reorganized as Liebig's Extract of Meat Company. Extract production was rather fitful through the early 1870s before Liebig and Giebert died within a year of each other in 1873 and 1874 respectively. Nevertheless, a huge market already existed for beef extract. By 1868, Liebig's Extract of Meat Company had sales agencies in nearly every European country, United States, Mexico, Venezuela, Brazil, Chile, and Bavaria; and consignments were sent to Cuba, Jamaica, India, China, Iraq, Turkey, and Estonia. During the early 1870s, the price of extract in Britain fell from two shillings to under one shilling per ounce.[4]

Relying on the ample supplies of native cattle that it acquired from drovers, as well as a small portion from its own *estancias* from the surrounding Rio de la Plata and Rio Grande do Sul regions, Liebig's typically used *criollo* oxen between four and six years old to make extract; it was believed that such cattle produced a darker and milder flavor of extract. One full-grown ox produced between eight and ten pounds of extract. That is, approximately thirty-four pounds of cattle flesh produced one pound of extract. Initially, Europeans consumed the extract in the form of a beef tea. One ounce of extract was boiled in three pints of water. Liebig's extract was cheaper than using actual cuts of beef and was free of fat and gelatine. Used as a health restorative and curative among all classes of Europeans, it was also used widely in hospitals and militaries. The other major use of the extract was as a spread, either alone or mixed with butter, for bread.[5]

Because demand steadily increased, production of extract at the Fray Bentos facility tripled between 1874 and 1891. Until 1889, Liebig Extract of Meat Company had a virtual monopoly on the product as well. During this period, Liebig's produced only a narrow range of meat products, aimed primarily at the growing meat demand among poorer people who wanted to consume more animal protein, specifically beef, which was considered the most valuable meat throughout Europe. One of these products was canned beef tongue; first produced in 1880, its very high fat content contributed to its reputation for flavor. Another was corned beef, launched in 1879, and arguably the company's most renowned product long into the twentieth century. A form of salt-cured beef, corned beef dated back hundreds of years in many parts of the world with the "corn" a reference to grains of salt. Corned beef was one of many industrial meat products that were made in Ireland in the early modern era and sold especially in England and France. While nearly 90 percent of the Uruguayan cattle that the company killed went for extract production, about 10 percent were used for corned beef. Starting in 1881, Liebig's pursued vertical integration by branching into cattle-raising on its own *estancias*.[6]

Liebig's Extract of Meat Company promoted its products around the world in various ways, including cookbooks. Liebig Company's *Practical Cookery Book*, published in England in 1893, for instance, stressed in the introduction that its product was produced in the cleanest conditions "under the strict supervision and control of the Company's chemist, who testifies as to the quality of meat used, and analyses and examines every parcel of Extract before its shipment to Europe." The introduction went on to caution consumers that competitors' products were "of an unreliable and even objectionable character. The Company's Extract is the only kind ever guaranteed genuine by Baron Justus von Liebig, and to protect themselves, the public should see that every jar purchased bears the facsimile of Justus von Liebig's signature, in blue, across the label." The cookbook offered 180 recipes in which extract was an ingredient and reminded readers that it could be used for beef tea and "for strengthening and improving the flavour of soups, sauces, gravies."[7] Nevertheless, as early as the 1870s and 1880s, many critics noted that extract had relatively little nutritional value. However, it did create a growing taste for beef among people who had had little access to beef before.[8]

Liebig's continued to stress production of meat items for poorer consumers worldwide long into the twentieth century. Although the company experimented with small-scale production of boiled and corned mutton, boiled beef, and potted tongue and other meats, its most important new product line at the start of the century was Oxo, a kind of beef extract that was both inferior to and cheaper than standard beef extract. Production began in 1900, and a year later an even cheaper extract, called Bulk Extract, was developed and sold to meat-potting firms. Unlike beef extract, which was a spreadable product, Oxo was a cube of beef stock that could be dissolved in water.[9]

In 1905, as global demand for inexpensive beef products continued to rise, Liebig's established another meat factory at Colón, Entre Rios, Argentina, about fifty miles further up the River Uruguay from Fray Bentos. Although worldwide demand for beef extract did not peak until about 1907, Liebig's began to shift more production to corned beef. More consumers wanted actual beef products instead of condensed beef. This transition narrowed the company's only other by-product production to canned beef tongues, which were produced in a Montevideo, Uruguay, factory. By 1910, corned beef had displaced beef extract as the company's primary product at both the Fray Bentos and Colón plants. The British Army's demand for corned beef in World War I, particularly one-pound cans suitable for soldiers' rations, pushed Liebig's to devote even more production to that product over beef extract, and shortly after the end of the war, the company decided to build a fresh meat production plant, a *frigorífico*, which opened at Fray Bentos in 1922. A trade dispute over importation of Argentinian cattle into Uruguay forced Liebig's to close the Frey Bentos plant in 1924. Leibig's had already shifted its extract production to a plant at Zeballos Cué, Paraguay, the year before. The British meat company Vestey then bought the Liebig *frigorífico* at Fray Bentos as part of its buildup of production capacity for the European frozen meat trade in the 1920s. Significantly, by that decade, Liebig's sales in Argentina outpaced its export sales.[10]

## Animal Protein and Human Health

As low-cost beef products garnered an increasingly larger consumer market in the West, widespread consensus existed within the medical and scientific community about the necessity of animal protein for human health. Liebig's efforts to make animal protein more widely available were complemented by efforts in the same era to determine more precise quantities of protein needed for human diets. Considerable debate occurred from the late nineteenth through the mid-twentieth century about how much protein was best for physical development. One of Liebig's students, Carl von Voit (1831–1908), supported his mentor's position on the importance of animal protein, though he and Liebig debated specifics related to the importance of albuminoids—proteins found in connective or supporting tissues—in muscular activity. In 1875, Voit proposed a daily dietary standard of 118 grams of protein, with at least half of the protein from animal sources, as necessary for the average physically active man. This amount—termed the Voit standard—quickly gained acceptance as the norm. The work of one of Voit's students, Max Rubner (1854–1932), on digestibility of proteins reinforced the Voit standard. In addition, the Voit standard was upheld by the American Wilbur O. Atwater (1844–1907), who received his PhD at Yale University in 1869 and studied in Germany with Voit and Rubner. Both Rubner and Atwater developed standard caloric values for basic food items, and both were interested in "rationalizing" food consumption. Rubner, Atwater, and many others were concerned about improving working people's nutritional health while at the same time economizing their food costs in the face of changes brought on by urbanization and industrialization. In particular, Atwater saw food as essentially a human power and energy source and wanted to divorce its consumption from specific and highly varied food cultures. He believed quantitative measures of food would make feeding the poor easier and cheaper. Rubner also wanted to make feeding the poor more efficient, but he feared the changes in diet brought on by urbanization; meat, wheat bread, sugar, vegetables, fruits, and dairy products had supplanted legumes, rye bread, and potatoes.

Rubner urged the German government to take measures that would improve the diets of working people. For instance, he wanted the state to establish a central food office that would study nutrition, establish workers' restaurants to provide cheaper and higher quality food, educate young people about nutrition, and create school lunch programs.[11]

While professor of chemistry at Wesleyan University from 1874 to 1907, Atwater was instrumental in shaping US government food policies. The first director of the USDA Office of Experiment Studies from 1888 to 1890, he also participated in the first dietary survey conducted in the United States in 1885 as part of a Massachusetts Bureau of Labor Statistics investigation. His research was then incorporated into a series of articles he wrote for *The Century* in 1887–88 on the importance of nutrition studies. The Massachusetts dietary studies indirectly examined laboring people's protein and caloric intake (the calorie was first used as a measure of food energy in 1883) through observation of food eaten and laboratory analysis of food samples. Starting in 1896, Atwater and Wesleyan physics professor Edward Rosa developed a more direct method of measuring caloric intake with their experiments using a respiration calorimeter, a room-sized closed and controlled chamber in which subjects lived for several days and chemical reactions were recorded from body and room temperature changes. Based on his work from the 1880s until his death in 1907, Atwater believed that working people's diets could be improved by making them more focused on cost-efficient nutritional content. As historian Nick Cullather stresses, Atwater embraced a scientific management perspective on nutrition and diet. Improvements in diet meant food choices based on cost instead of taste or prestige. His research underscored efforts of industrialists, philanthropists, and home economists who argued that raising wages for workers was not necessary if workers simply made more cost-effective food choices. Atwater's ideas were also central to home economists' efforts to Americanize immigrants' diets.[12]

Atwater's arguments about a moral economy of food and the need for workers to rationalize and economize their food choices are evident in *Food and Diet*, published by the USDA in 1895. This report drew partly upon the dietary survey of the Massachusetts Bureau of Labor Statistics, as well as European studies of workers' diets. It argued that food choices should be made according to food's protein and caloric content cross-tabulated by cost. Atwater was outspoken about workers' tendencies to be wasteful in terms of food choices. He asserted that the chief problem in Americans' food economy was that people "purchase needlessly expensive kinds of food."[13] He dismissed social or cultural contexts for food choices and denigrated purchases of prestige foods, like beef tenderloin, which for workers symbolized participation in a more prosperous way of life. Labor union leaders such as Eugene V. Debs criticized Atwater's views as attacks on workers' efforts to improve their standards of living. Werner Sombart's 1906 classic *Why Is There No Socialism in America?* noted that American workers ate more meat of all kinds, as well as more eggs, potatoes, fruits, and vegetables than British workers did. Labor leaders and social reformers wanted to increase the quantity and quality of food, not economize it.[14]

While Atwater's focus was primarily aimed at urging working-class Americans to be more mindful of cost efficiencies in their food choices, his publications also underscored the importance of eating meat. Because working people had to spend relatively more of their income on food while at the same time expending more energy than middle- and upper-class Americans, he urged them to eat cheaper types of meat. Atwater's tables compare wide varieties of food for their relative costs. "Meats, fish, poultry, and the like are expensive, while flour and potatoes are cheap food. The reason for this is simple. The animal foods are made from vegetable products. Making meat from grass or grain is costly."[15] Instead of dispensing with meat, Atwater urged Americans to choose "the lean of meat and fish and the gluten of wheat."[16] His tables encouraged the thrifty-minded to purchase salted and canned meats and fish over fresh mutton, pork, and beef.[17]

Liebig, Voit, Rubner, and Atwater all asserted that relatively large amounts of protein were necessary to maintain normal physical activity. For example, based on the Massachusetts dietary survey, Atwater found that average protein consumption was 126 grams with 52 percent from animal sources. Russell Chittenden, often referred to as the father of American biochemistry, challenged this standard and argued that one-half that amount could maintain normal physical activity. Chittenden received his PhD in chemistry from Yale in 1880 and then served as a professor of physiological chemistry there from 1882 to 1922. *Physiological Economy in Nutrition* was published in 1904 and is generally considered his most important work on protein standards. The book described his experiments on himself, four colleagues, eight college students, and thirteen US Army volunteers, and made the case that the Voit protein and caloric standards were far too high. The Voit caloric standards ranged from 2,200 calories for sedentary people all the way to 4,500 calories for very strenuous activity levels. Chittenden believed that basic human physical functions could be achieved with far fewer than 3,000 calories per day. His protein calculations have proven to be enduring and well accepted; current USDA Dietary Guidelines stipulate 46 grams for women and 56 grams for men. However, his caloric estimations—2,100 calories for moderately active men and 2,500 calories for more active men—are now considered somewhat too low for all but completely sedentary people. Current USDA guidelines indicate a range of 1,800 to 2,200 calories for middle-aged women and 2,400 to 3,000 calories for middle-aged men, depending on activity levels.[18]

Chittenden's *The Nutrition of Man*, based on eight lectures that he delivered on human nutrition in early 1907 at the Lowell Institute in Boston, encapsulated his work and made the case for new dietary standards while maintaining an argument consistent with Rubner and Atwater regarding the need for greater rationality and efficiency in food choices, including meat. "Much stress is ordinarily laid upon the importance of a large intake of proteid food whenever the body is called upon to perform severe, or long-continued, muscular work; but in view of what has been stated it may be questioned whether there is any real physiological justification for such custom."[19] Chittenden did not reject the need for animal protein since "[m]an is by choice an omnivorous creature."[20] Instead, he emphasized the need to consume other forms of protein: "we must recognize the value of the non-nitrogenous foods as a source of energy, and this means plainly food from the plant kingdom."[21] He reinforced Atwater's emphasis about the public's need for better understanding of food, health, and economy, but also pointed out that Atwater's focus on "lean meat, eggs, fish, etc." was not enough.[22] Chittenden urged people to consume dried peas and beans, almonds, pine nuts, and cheese as sources of protein and to be mindful that simplicity and moderation should guide food choices.[23]

## Meat and the Eugenics Debates

Even as more middle-class and, especially, working-class Westerners availed themselves of cheaper meat products, the vegetarian movements in Britain and the United States, which began in the early nineteenth century, attracted increasing interest and adherents from the educated middle and professional classes. Beginning with so-called Bible Christians like William Cowherd, William Metcalfe, and, most notably, Sylvester Graham, the movement to abstain from eating meat gained more acceptance after the founding of the British Vegetarian Society in 1847, followed by the American Vegetarian Society in 1850. The early leaders of both the British and American Vegetarian Societies were generally involved with a variety of social reform movements, including abolitionism, temperance, spiritualism, and radicalism. Importantly, in both Britain and the United States, the vegetarian movements' ideals and concerns represented ethical convictions that connected meat eating to violence and aggression toward both humans and animals. Over the last half of the nineteenth century and well into the twentieth, vegetarians pursued a nutritional campaign stressing ethical and food safety themes against the growing medical and scientific literature as well as the dominant food cultures that stressed that more meat was necessary to human health and well-being. Although the vegetarian side of the debate over meat's role was never dominant during this era of meat promotions, its positions both sharpened its opponents' emphasis on meat's importance and also laid the groundwork for a resurgent antimeat campaign in the West during the post–World War II era.[24]

Although vegetarians generally emphasized moral arguments to support their views about human health, some also buttressed their claims by using science. John Harvey Kellogg likely did more than any other vegetarian in this era to wield science in the cause of meat avoidance. In 1876, after obtaining his medical degree and training at the University of Michigan and the Bellevue Hospital Medical College in New York City, Kellogg became medical director of the Battle Creek Sanitarium in Battle Creek, Michigan. Until his death in 1943, Kellogg directed the San, as it was popularly known, and made it into the largest health facility of its kind in the United States. Inspired by his Seventh Day Adventist antimeat beliefs, Kellogg tirelessly promoted vegetarian diets through the San's activities as well as a variety of publications including the San's monthly publication, *Good Health*. One of Kellogg's emphases was that, since meat was high in nitrogen, it would in moist conditions, such as the human large intestine and colon, putrefy and turn alkaline, which in turn would result in autointoxication. Kellogg was one of many physicians in this period who held such views. Current medical thinking stresses that autointoxication, or autotoxicosis, stemming from decomposition of intestinal waste products, contributes to acidosis—high acidity not alkalinity—which can cause colon cancer.[25]

A prolific author of about sixty books and hundreds of essays and pamphlets, nearly all of which emphasized vegetarianism, Kellogg's most focused attack on meat eating was his 1899 book titled *Shall We Slay to Eat?* In it, Kellogg made the case that human beings' physiology favored a vegetarian diet. He focused considerable attention, as was often the case in this era's vegetarian writings, on the parasites and diseases that people acquired from eating meat. He singled out canned-meat products as the most common source of meat poisoning. In addition to grounding his remarks in the science of meat's dangers, Kellogg also cited Biblical authority and ethical arguments to warn his readers. Using a literal reading of the Bible, he noted that several Old Testament figures were not only vegetarians but also lived for hundreds of years. He pointed out that Noah, for example, lived to be 950 years old. Kellogg used Christian teachings to chide meat-eaters as egotistical pleasure-seekers: "Flesh eating is but the natural result of that supreme selfishness that leads man to the egotistical belief that all things were made for his own personal pleasure and use." Indeed, Kellogg's ethical arguments were made powerfully compelling when he referred to humans' ties as sentient creatures to all living animals: "The basis for the ethical argument against flesh eating is to be found in the fact that lower animals are, in common with man, sentient creatures."[26] Kellogg expressed outrage against the horrors of the modern slaughterhouse—seven years before publication of Upton Sinclair's *The Jungle*—as he described Chicago's Union Stockyards and the meatpacking plants found there: "What crime have these poor brutes committed that they should thus be executed? What law of God or man have they violated that they should thus prematurely die, that their blood should be poured out upon the soil as a fertilizer? Verily, the blood of multimillions of innocents cries from the ground."[27]

Prominent vegetarians also supported the eugenics movement. Widespread among medical authorities, professionals, and academics in the early twentieth century, the eugenics movement argued that national welfare depended on the hereditary traits of its people and that science needed to improve those traits to develop stronger societies. Irving Fisher (1867–1947), professor of political economy at Yale University, was one of the most prominent academic proponents of a vegetarian-infused eugenics during the early twentieth century. A widely published scholar in various areas of economics, his brush with death from tuberculosis in 1898, which kept him out of teaching for three years, propelled new interests in diet and endurance. In his 1908 book, *The Influence of*

*Flesh-Eating on Endurance*, he described an endurance experiment in which forty-nine people represented three groups: athletes who ate high protein diets; athletes who ate low protein diets; and sedentary people who ate low protein diets. Comparing their performances in endurance exercises, he drew the conclusion that significant meat-eaters showed less endurance than vegetarians, including those who were sedentary. He attributed the difference to meat eating's production of more uric acid, which, he noted, was in line with Russell Chittenden's theories. A by-product from high levels of purine consumption from meat and sugar, uric acid is also considered by medical authorities today to be a cause of gout, kidney stones, and, possibly, hypertension. In 1908, Fisher and Horace Fletcher founded the Health and Efficiency League of America, which, though short-lived, was an attempt to build support for a federal cabinet-level Department of Health.[28]

Fisher's interests in various health causes, including promotion of a Life Extension Institute and of prohibition, propelled his involvement in the eugenics movement, which also included John Harvey Kellogg among its adherents. In the context of the movement's racism, dietary reform was believed to be central to strengthening and protecting white America against both race suicide and the threat of immigration. Beginning in 1914, Kellogg sponsored Race Betterment Conferences held at the San that attracted a variety of reformers interested in eugenics, human biology, and immigration issues. Fisher was part of the inaugural conference and then part of coalition of eugenicists who attempted a so-called eugenics peace initiative among adherents in Britain, France, Germany, and the United States to formulate strategies about how to protect the white race against the "threat" of Asian immigration. He was then one of the leading American eugenicists who petitioned the federal government to implement eugenic proposals.[29] In a 1921 address as president of the American Eugenics Research Association, Fisher blamed various problems of civilization that he felt had undermined the health and future well-being of Americans: "Civilization has thrown the daily life of the individual out of balance, so that not one person in a hundred lives what might be called a biologic life. He is insufficiently exposed to the air, he eats too fast and often too much. In America he eats far too much protein and far too little bulk. His food is far too soft. It is usually lacking in vitamines [*sic*]."[30] Fisher was then one of the three American eugenicists present at the founding of the International Federation of Eugenics Organizations in 1925.[31]

Parallel to the leading American vegetarian eugenicists, meat advocates expressed a version of eugenics that espoused the importance of animal protein. During the first two decades of the twentieth century, Woods Hutchinson (1862–1930), an Englishman who received his medical degree from the University of Michigan and then taught at various American universities from the 1890s until his death in 1930, was one of the most popular advocates for the importance of meat in human diets during the era. Much of his advice about proper diets, especially as found in his *Handbook of Health*, first published in 1911 and then reprinted in many editions, directly challenged the positions and advice of vegetarian eugenic advocates, particularly John Harvey Kellogg. Indeed, in a 1919 editorial in *Good Health*, the official publication of the San, Kellogg described Hutchinson as "a beef drunkard and propagandist of the most pronounced type."[32]

In Hutchinson's *Handbook of Health*, which provided health advice on a wide variety of topics beyond diet, he started with a classification of foods under the general description of the "Food-Fuel of the Body-Engine." He categorized foods in descending order of importance as coal foods, kindling foods, and paper foods. The coal foods—that is, foods that provide energy and strength—include proteins, starches-sugars, and fats, which make them staples. He claimed that proteins are "the First Foods," and also the "most necessary foods." Although expensive and sometimes dangerous when allowed to become tainted or spoiled, "taken all together, the proteins, or meats, [were] the most nutritious and wholesome single class of foods."[33] Unlike John Harvey Kellogg's food advice in *The New Dietetics: What to Eat and How: A Guide to Scientific Feeding in Health and Disease*, which waxed eloquently on the virtues of starchy root vegetables, particularly potatoes, as well as legumes and fruits, Hutchinson had very little to say about any of these foods. Nearly all vegetables and fruits were treated somewhat dismissively as lacking in food value and wholesomeness in the single chapter devoted to "Kindling and Paper Foods—Fruits and Vegetables." Regarding nuts, particularly peanuts, which Kellogg's *New Dietetics* highlights as part of the legume family like beans and peas, Kellogg emphasized that "evidence is constantly accumulating to show [peanuts'] great food value,"[34] whereas Hutchinson mentioned only in passing their value as a fat but noted also that they are "likely to upset the digestion."[35] He said nothing about Kellogg's efforts to address peanuts' digestibility through preparation of peanut butter, a food product made famous at the San.

With regard to the potato, which Kellogg was enamored with—noting, for instance, that "if the average American would eat four or five times as much potatoes as he now eats, and an equivalent amount less of cereals and cane sugar, the result would be a notable lowering of the death rate and a marked increase in physical fitness and freedom from crippling disease"[36]—Hutchinson conceded that "the only important starchy food outside of the grains is potatoes." Hutchinson's most direct jab at Kellogg was in regard to breakfast cereal. He described cereal as mostly made with the "waste of flouring mills" and lacking in food value. "They should never be relied upon as the main part of a meal."[37]

His focus on wholesomeness also betrayed Hutchinson's eugenics perspectives that challenged much of Chittenden's and Kellogg's work. As he used the term "wholesomeness" in *A Handbook of Health*, Hutchinson meant food choices that reflected appetites derived from one's culture. In a manner not unlike those who currently advocate paleo diets, he asserted, "[F]rom the millions of years of experience that the human body has had in trying all sorts of foods, it has adapted itself to certain kinds of food and developed certain likes and dislikes which we call *appetites*" [italics in original].[38] In contrast, Chittenden's research had led him to conclude just the opposite: "Habits and cravings are certainly very unreliable indices of true physiological requirements." In fact, Chittenden believed it was relative affluence that contributed to Westerners' predilection for a meat-rich diet.[39] Hutchinson explained his cultural perspectives more thoroughly in his 1908 book, *Instinct and Health*. In it, he connected food choices to culture and made the case, in a manner strongly suggesting a Social Darwinist influence, that food choices stemmed from cultures that were either relatively superior or inferior to each other. He described Jewish and Muslim traditions that opposed pork consumption as "primitive prejudices . . . [that] are traceable to the belief that the animal is the totem, or guardian spirit, or primitive ancestor of the tribe," whereas for Christian cultures "[p]ork, including ham and bacon, is easily our second most valuable meat food [behind beef], and has laid the literal foundation of our Western Civilization." He went on to generalize that vegetarianism's foundation "is *that the eating of animal food is morally wrong*" [italics in original]. But since vegetarianism's adherents viewed their dietary choices as akin to religious choices they were inherently weak: "With vegetarianism as a creed, we have, as scientific men, no more quarrel than with any other creed. But when it parades in the guise of science we firmly but respectfully protest."[40]

Hutchinson's dietary advice was intertwined with his eugenics perspectives, and he was quick to connect vegetarianism to weak and uncivilized cultural traditions. This dietary advice underscores his support for negative eugenics, that is, the elimination of aspects of cultural inheritance that make a population weak. For instance, in *Instinct and Health*, he said that "it may be stated that vegetarianism is the diet of the enslaved, stagnant, and conquered races, and a diet rich in meat is that of the progressive, the dominant, and the conquering strains."[41] In *A Handbook of Health*, his message was a bit less racist but no less clear: "Those nations which live solely, or even chiefly, upon one or two kinds of staple foods, such as rice, potatoes, corn-meal, or yams, do so solely because they are too poor to afford other kinds of food, or too lazy, or too uncivilized, to get them; and instead of being healthier and longer-lived than civilized races, they are much more subject to disease and live only about half as long."[42] Regarding the vegetarian's concern for animal life, and even the horrors of vivisection, Hutchinson was quick to assure his audience that sacrifice of animal life was natural and necessary for civilization. As he explained in *Civilization and Health*, published in 1914, "Our very lives depend upon the meat and milk and skins of cattle, upon the wool and flesh of sheep, upon the products of the poultry yard, of the rabbit-hutch and pigeon-cote." Later in the same chapter, he maintains, "All things that live must die, and the utmost that we ask for ourselves as men is that we shall not have died in vain."[43]

Hutchinson dismissed vegetarian food choices as both unnatural and uncivilized. He believed that vegetarian food choices flew in the face of civilized traditions that had made "coal foods" wholesome. In contrast, John Harvey Kellogg and Russell Chittenden saw health value in East Asian food traditions. Regarding soy, which Hutchinson did not mention in *A Handbook of Health*, Kellogg says in *The New Dietetics* that "soy fills the place of meat as well as milk in the dietary of many millions of sturdy Orientals."[44] In *The Nutrition of Man*, Chittenden contextualized his remarks on the healthfulness of a relatively limited protein intake by citing Ōshima Kintaro's work that was published in 1905 by the USDA as *A Digest of Japanese Investigations on the Nutrition of Man*. Although Chittenden was clear that he believed the Japanese people would benefit from a mixed diet of vegetables and meat, he nonetheless quoted Ōshima: "It is sometimes remarked . . . that the peasants in the rural districts of Japan, living largely on vegetable food, are really healthier and stronger than people of the better classes, who live on a mixed diet, and the better physical condition of the former is commonly believed to be due to their diet." Chittenden commented that "[d]oubtless a mixed diet, with a larger proportion of animal food, did their means readily permit, would offer some advantages from the standpoint of palatability and variety, but it is questionable if any material gain in health or strength would result."[45]

## Promoting Animal Protein in East Asia

Russell Chittenden's assertion that "[h]abits and cravings are certainly very unreliable indices of true physiological requirements" is notable in the context of the promotion of animal protein in East Asia during the late nineteenth and early twentieth centuries. European observers of both China and Japan during the mid- to late-nineteenth century generally found the native food cultures to be both distasteful and lacking in nutrition, especially in terms of meat. Many of these observers were imperialists and racists, such as English physician Patrick Manson (1844–1922), who spent his early career in China before returning to England to eventually lead the London School of Tropical Medicine. About the near absence of meat in the Chinese diet, Manson remarked that "[t]he savage races exist by adapting themselves to circumstances, the more highly civilized races by adopting circumstances to themselves."[46] Over the course of this era, the Japanese in particular and the Chinese as well, though to a lesser degree, struggled to create habits and cravings where they had not existed before and, in doing so, also addressed the charge of Patrick Manson in terms of increasingly adopting circumstances to themselves. Both the science that advocated meat eating in the West and the commercial promotions of meat extract made inroads in China and Japan.[47]

As noted, the Japanese had eaten meat for medicinal purposes for some time before the late nineteenth century, but it was nevertheless negligible in most people's diets. The Emperor Meiji's lifting of the long-standing ban on meat consumption in January 1872 was symbolically important but had had little impact on general consumption. Perhaps the first significant incorporation of meat among the Japanese people occurred in the navy and army. The disease that the Japanese called *kakké,* which was later found to be beriberi, was endemic among the Japanese middling ranks. As now understood, beriberi is caused by a lack of vitamin B-1, or thiamine, which is found in whole grains, meat, fish, eggs, dairy, green leafy vegetables, and legumes. Takaki Kanehiro, a young surgeon who joined the Japanese navy in 1870, spent much of his military career studying the disease. He rejected the widespread belief among both Japanese and Westerners that it was caused by miasmas from wet soils because the patients he saw were ship bound. His work focused on nitrogen in sailors' diets and seems to have built on the work of Westerners, perhaps especially Van Leent, a Dutch naval surgeon, who in 1873 fed fresh eggs and coconut oil to Javanese sailors with beriberi. By 1883, Takaki had discovered that the nitrogen to carbon ratio in the diets of Japanese sailors was lower than that of European sailors and was also lower than the ratio found in Japanese naval officers' diets. Common sailors had a much higher incidence of beriberi than European sailors and Japanese officers. In that year, he convinced naval authorities to modify sailors' meals to include greater proportions of meat, condensed milk, bread, and vegetables than rice. Once this was accomplished, the incidence of beriberi declined sharply. The Food Supply Act of 1884 made meat part of naval rations, and by 1887, all naval ships and depots made these modified meals available. The army first introduced canned beef in 1877, but its rations followed the path laid out by the navy's changes.[48] Historian Katarzyna J. Cwiertka argues that by the early twentieth century Japanese soldiers' diets were far superior to the average Japanese diet and that this diet "contributed to the moulding of a nationally homogenous taste."[49]

While military service introduced meat to one segment of the Japanese population, the introduction of Liebig's extract helped to introduce other segments of both the Chinese and Japanese populations to meat's taste and nutritional benefits. The product was available in China as early as 1868. Shortly after Isabella Bird's 1878 account of food availability that suggested that Liebig's extract was not yet available, Liebig issued trading cards of its products in Japan as early as the 1880s. The widespread distribution of Liebig product trading cards in both China and Japan during the 1880s suggests that sales of meat extract became more popular in East Asia during the late nineteenth century.[50]

The last quarter of the nineteenth century saw increasing efforts among Westerners in East Asia as well as Japanese medical authorities to promote meat. The Meiji government modeled much of its public health measures after Western ideas and techniques. As noted in chapter 2, Fukuzawa Yukichi was the earliest prominent Japanese intellectual to support meat consumption. His book, *A Treatise on Meat Eating*, promoted meat consumption as a means of both encouraging physical vitality and building national strength. During the 1870s and 1880s, European physicians arrived in Japan to provide guidance to the country's newly established medical universities. Many of them focused considerable attention on Japanese diets and nutritional standards. Nevertheless, many, including, for instance, Erwin Baelz, a professor at Tokyo Imperial University, surprised themselves with the discovery that Japanese nutritional intake was already adequate. Still, several Japanese medical authorities emphasized the need for their people to eat more meat. Tokyo Imperial University professor of physiology Ōzawa Kenji connected meat to national strength as did Tōan Matsumoto, the first principal of Keiō University's Medical School. Both adopted a Japanese version of the eugenics message when they stressed that Japan faced racial extinction unless people ate more meat. During the same era, prominent Japanese philosophers made the same case. Katō Hiroyuki and Inoue Tetsujirō worried that Japanese people were weaker physically and mentally than Westerners and needed to eat more meat in order to catch up. Takahashi Yoshio's *The Improvement of the Japanese Race*, published in 1884, strongly advocated meat eating as essential to improving the Japanese people's health.[51]

To be sure, by the early 1920s, China and Japan were a long way from being significant meat-consuming countries. In Japan, some members of the elite, the educated middle class, and segments of the military had become occasional meat eaters. Few Japanese commentators at the beginning of the twentieth century believed that Japan could view meat eating as a practical means of increasing strength as Ōkuma Shigenobu had advocated before he became prime minister in 1898.[52] Two decades later, in 1921, another Japanese commentator noted that the obstacles to the country's ability to produce enough livestock for meat consumption, particularly availability of land, were too great. "It is, however, a matter of great regret that the special circumstances found in Japan which have been referred to will always prevent live stock [*sic*] farming from becoming a very strong industry."[53] Robert James McFall's *The World's Meat*, published in 1927, confirmed this assessment, though he noted that the government "is doing its utmost to stimulate the live-stock [*sic*] industry and is already setting an example to other oriental lands in the aid which it is giving in improving the native stock and stamping out disease."[54] McFall noted that while China's peasants raised livestock for traction power and manure and that some meat was exported to the Philippines and Japan, its domestic consumption was still limited. Meat would not contribute significantly to the diets of most Japanese and Chinese people until the latter half of the twentieth century.[55]

Before World War II, the first food regime relied partly on an international meat trade that reinforced meat-eating cultures in the West. Japan and China fell outside this meat trade. As explained in chapter 1, the first food regime's meat trade created increasing availability and demand for meat through the development of Western settler nations, where meat consumption was high and surplus meat was shipped back to the West, especially to the UK, for its increasingly meat-hungry population. While Japan and China attempted to develop their livestock industries as well as meat eating during the late nineteenth and early twentieth centuries, the results were limited before World War II. Chapter 6 points out that the key to both nations' creation of greater meat availability was participation in the second food regime, in which the Americas provided the feed grains necessary for them to focus on development of their own livestock production and meat industries, supplemented by increasing meat trade with the West. Chapter 4 will provide evidence and examples of the crucial role of the state in developing meat industries, initially in the West and then in East Asia.

# 4

## The State and Meat

Beginning in the late nineteenth century, governments in the West and many non-Western countries established regulatory practices that encouraged meat production and consumption. Two major and related concerns influenced these developments: increasing meat trade and satisfying consumers' growing demand for wholesome meat products. Governments attempted to achieve these goals on multifaceted fronts. Perhaps most central were efforts to control animal diseases and meat parasites. Consumers around the globe in the nineteenth century were especially concerned about the hygienic quality of the meat they ate. Increased meat consumption and trade required greater attention to meat safety. Concerns about increasing meat consumption and trade were often intertwined, as in the so-called German-American Pork War from 1880 to 1891.

Control of production was another focus of state regulatory involvement in meat production and consumption. In US history, meatpacking, just as with nearly all other industries, has almost always fallen within the sphere of private enterprise, though some advocated for publicly controlled meat production, especially during the Progressive Era. Although the character of private ownership varied depending on the size of the firm, type of ownership, and national versus foreign ownership, the fact that the United States has no history of public ownership of meat production is significant in many ways that can be understood only in a global context. Throughout Western Europe and East Asia, public, usually municipal, ownership of meat production became commonplace in the late nineteenth and early twentieth centuries. The variation among countries in terms of private versus public meat production reflected not only underlying differences in economic practices but also differences in concerns about problems especially associated with meat production's health and environmental effects. To a lesser degree, public ownership also involved regulation of work conditions, as explained in chapter 7. While business practices varied, increased provision of meat for consumers was the goal. Not until the late twentieth century would national governments become involved in efforts to inform their citizens about food choices in ways that questioned the desirability of meat consumption. This chapter highlights the ways in which governmental involvement in meat production and consumption increased consumer safety and influenced meat production's business structures and workplace environments and, over the course of time, contributed fundamentally to making meat an important part of human diets around the globe.

## Controlling Animal Diseases and Meat Parasites

Concerns about animal diseases that killed livestock, called epizootics, and livestock diseases that might lead to human illness, known as zoonoses, were longstanding in European history. Significant state intervention to prevent transfer of disease from animals to humans occurred in Western Europe especially during the rapid rise of meat consumption that started in the eighteenth century and accelerated during the nineteenth century. Estimates of the magnitude of diseased meat became much more reliable by the mid-nineteenth century. Surgeon and former veterinary student Joseph Sampson Gamgee estimated that 20 percent of all the meat sold in Britain in the 1850s came from diseased animals. In 1862, Britain's Privy Council confirmed this figure. Two epizootic diseases, pleuropneumonia, a contagious bacterial disease that affects the lungs of cattle, and rinderpest, a deadly cattle virus, caused considerable panic about animal disease during the mid-nineteenth century. However, the focus of concern in Britain about animal disease shifted to a zoonotic disease, bovine tuberculosis, by the 1870s and 1880s.[1]

Bovine tuberculosis results from infection by a bacterium, *Mycobacterium bovis*. It had been identified as a zoonotic disease in British medical publications as early as 1876. Robert Koch's discovery of the tubercle bacillus in 1881 helped to convince scientists that bovine and human tuberculosis were clearly linked. By the mid-nineteenth century, Britain's increasing urbanization, transportation of animals to urban markets, and greater consumption of meat and milk all contributed to the disease's rising incidence. It was also problematic because the disease often took months or years to incubate before any obvious symptoms in cattle were evident. Yet, especially in young children, the disease could be devastating. In 1908, heart disease and tuberculosis were the two leading causes of death in England and Wales among men. While medical understanding of the disease's transmission increased dramatically after the 1870s, many among the British public, including large segments of those who worked in the livestock and butchering trade, viewed the problem as one caused by unhygienic living conditions among the urban poor. Certainly the urban poor, whose consumption of meat and milk rose rapidly after the mid-nineteenth century, were most susceptible to poor quality meat, often purchased in the form of sausage or convenience meat items. Many British consumers also preferred to eat undercooked meat despite widespread advice in the medical press—starting in the 1860s—that it was dangerous to do so.[2]

Medical understanding of the health effect of bovine tuberculosis kept just ahead of efforts to protect the meat-eating public through the end of the nineteenth century. Until the nineteenth century, efforts in England to control diseased meat, primarily undertaken by butchers' organizations and guilds, had been haphazard at best. Passage of the Nuisances Removal Acts of 1846–48 provided support for better inspection, while the Public Health Act of 1848, Nuisances Removal Acts of 1855 and 1863, and the Adulteration Acts of 1860, 1862, and 1872 were also helpful. Perhaps the most important legislation supporting inspection was the Public Health Act of 1875. By the mid-1880s, local public health officials regularly examined meat. Unfortunately, however, these sanitary inspectors were typically un- or underqualified, and veterinary doctors were often excluded from the inspection process. Moreover, butchers' organizations, particularly the National Federation of Butchers and Meat Traders' Association, established in 1888, often actively interfered with the inspection process and defended butchers against what they saw as unnecessary persecution of members of their profession. This antagonism between sanitary inspectors and butchers was one of the reasons why many medical and veterinary groups, as well as activists promoting humane slaughter, joined together to support establishment of public slaughterhouses. Scotland moved to replace private with public slaughterhouses by the end of the nineteenth century. By 1910, 60 percent of Scotland's cities had public slaughterhouses; and, by 1930, this figure reached 80 percent. In contrast, England, as explained further below, was only slowly moving in this direction by the World War I era.[3]

Because of the uneven quality of inspection, after the mid-1890s greater efforts were made to eradicate bovine tuberculosis. By 1914, a standard examination for the disease was available as well as significant supporting medical and veterinary literature. During the early twentieth century, much of the fear of tuberculosis shifted from meat to milk. Part of the problem was that bovine tuberculosis, unlike epizootics such as rinderpest, pleuropneumonia, and foot-and-mouth disease, had not been included in the 1869 Contagious Diseases (Animals) Act. Many farmers continued to argue that bovine tuberculosis was hereditary, though by World War I there was no serious medical disagreement that it was a zoonotic disease. Unlike pleuropneumonia, which was eradicated in Britain by 1898, bovine tuberculosis was not eliminated until the 1960s in Britain. In the United States, where one veterinary expert in 1908 said as many as one-half of all dairy cows might be infected and where one in five children under five and one in ten adults were infected with the disease, the slow pace of both inspection and eradication made pasteurization of milk the major means of controlling the impact of bovine tuberculosis. Although some medical experts lamented the effects of heat on milk's nutritional quality, pasteurization was largely effective in reducing the incidence of the disease.[4]

Attempts to eliminate the threat to cattle from rinderpest, a deadly virus that had ravaged cattle populations over centuries in Europe, set in motion a similar set of developments that led to increased state intervention in both France and Germany during the nineteenth century. Rinderpest was a virus that caused cattle to shiver and cough initially, followed by drooping head and blistered tongue, followed then by diarrhea that usually killed the animal within one to two weeks. It was particularly deadly in Western Europe and Britain during the eighteenth and nineteenth centuries. Between 1711 and 1799, it killed an estimated 200 million cattle in Europe and 28 million in the German states alone. A particularly devastating outbreak of the disease in France in the 1770s sparked the French government to form a commission to assess the problem. The commission's work led to the founding of the Société Royale de Médecine. This body disbanded in 1793 but became the forerunner of the Royal Medical Society, founded in 1820. Another rinderpest outbreak in Britain killed roughly three hundred thousand to four hundred thousand cattle between 1865 and 1867. Systematic state intervention throughout Western Europe and Britain in the 1800s eliminated rinderpest's threat in these areas by 1880.[5]

In 1802, Napoleon reorganized the municipal surveillance authorities over the butcher trade in Paris because of meat's strategic military importance. Then in 1810, he ordered the creation of five public slaughterhouses in the city, which were opened eight years later. The centralization and regulation that accompanied these new slaughterhouses to insure safe and plentiful meat supplies became major elements of the impetus for the public slaughterhouse movement throughout Europe and other parts of the world in the nineteenth and twentieth centuries. Germany likewise made both rinderpest control and public slaughterhouse development major focal points starting in 1869, when a new law was passed giving the state new powers to deal with animal disease outbreaks. One of the major steps of the new German state was the 1881 creation of the Central-Viehhof in Berlin, a public slaughterhouse that would allow better management of potential disease threats.[6]

In the United States, widespread concerns about several epizootic animal diseases, including pleuropneumonia and Texas fever in cattle starting in the mid-1860s, resulted in the formation of the Bureau of Animal Industry (BAI) in 1884. The BAI's efforts to eradicate the tick that carries the parasite *Babesia* and causes babesiosis, or Texas fever, were focused on increasing the value of the hides from southern cattle marketed in the Midwest. Arguably hog cholera, which was known also as classical swine fever, was the most significant animal disease that sparked the bureau's formation. Likely imported with pigs from Europe, since the disease was described in France in 1822, the first reports of hog cholera in the United States occurred in 1833. Animals infected with the disease initially exhibited a fever followed by fatigue, vomiting, and diarrhea, and difficulty breathing before death. Calculations made in 1917 estimated that direct losses for the preceding forty years due to hog cholera averaged forty million dollars per year and that the death rate was 40 to 103 per thousand. Its serious economic impact on the pork industry made the BAI's intervention essential.[7]

The BAI was instrumental in identifying the nature of hog cholera, producing a variety of serums, determining how the disease was transmitted, and working over decades to develop vaccines. The BAI's first director, Dr. D. E. Salmon, was involved in identifying *Bacterium suis*, later known as *Salmonella choleraesuis*, which often occurred with hog cholera. Although Salmon originally thought the disease was due to bacteria, by 1903 the cause of hog cholera was determined to be a virus. A serum to combat the virus was developed in 1906–07, and trials were held at the Omaha Stockyards beginning in 1910. Field experiments held in Iowa from 1912 to 1914 showed the effectiveness not only of the serum but of the BAI's strategy of encouraging cooperation among farmers, state officials, federal government authorities, and practicing veterinarians. These cooperative strategies were applied in 1916–18 to experiments in determining how the disease was transmitted and showed that many outbreaks of cholera were due to pigs eating infected meat scraps and meat-eating birds that carried the virus. Efforts to develop effective vaccines took much longer; experiments began in 1903 and resulted in successful vaccines only by the 1930s.[8]

Concern about the quality of American pork exports to Europe was the other major factor in encouraging the formation of the BAI, as well as the first significant meat inspection act in US history. By the mid-1870s, the United States produced nearly as many hogs as all of Europe: 32.2 million versus 43.5 million. Of the American total, 60 percent were exported. In 1879, the United States exported 1.2 billion pounds of pork products worth approximately $80 million. This amount constituted about 10 percent of total US exports. After a University of Vienna professor warned that as many as 20 percent of all American hams might be infected with *trichinae*, the German-American Pork War ensued and raged over the next decade. *Trichinella spiralis* is a parasitic nematode (roundworm) founded in warm-blooded mammals, including pigs. This health concern also coincided with German chancellor Otto von Bismarck's attempts to provide trade protection for animal production among his country's farmers in the face of American productive capacity and export interests. Other European nations were also protectionist for similar reasons.[9]

Although Italy, Portugal, Spain, Greece, France, Austria-Hungary, Turkey, Romania, and Denmark all participated in the ban on American pork products during this period (Britain banned live swine imports due to concerns about hog cholera but continued to import American pork products), Germany was the most notable in terms of its volume of trade overall with the United States. German meat dealers expressed considerable ire over the fact that American pork products were 40 percent cheaper than theirs. Although trichinosis was problematic in many areas of the world, German concerns were especially notable, partly because many people in different regions of the country, perhaps most notably in Saxony, ate raw pork; thorough cooking of pork otherwise killed the *trichinae*. Between 1877 and 1879, an average of thirty-three people died each year in Germany from trichinosis. German authorities had developed microscopic inspection procedures as early as 1863 to cope with the fact that undercooked pork consumption was a cultural preference that could not be easily eradicated. The stigma of diseased German pork even spilled over into the growing imperial antagonism between Britain and Germany during this period.[10]

The German government barred American pork products in two stages. On June 25, 1880, the government banned ground pork and sausage with passage of the Imperial Law Concerning the Prevention and Suppression of Animal Diseases; on March 6, 1883, all other pork products, particularly bacon and ham, were excluded. The government's actions set off a decade-long diplomatic and media furor in both countries over the pork trade. German authorities generally insisted that American authorities simply needed to institute microscopic investigatory procedures as it had, while American authorities claimed that protectionism fueled German actions. The US Department of State sponsored an investigation of the pork industries in Cincinnati and Chicago by Michael Scanlon, chief of the statistical bureau, and he concluded that trichinosis was not a significant problem. However, at the turn of the twentieth century, an estimated 2.5 percent of hogs were likely infected. Between 1901 and 1915, 240 Americans died from trichinosis. Postmortem surveys conducted in the 1930s and published in a 1943 National Institute of Health report suggested that 16.2 percent of Americans might be infected. Scanlon's likely politically motivated finding delayed American action to better police the health status of its meat, as did slow acceptance of scientific findings about the role of microorganisms in disease and the reluctance of the federal government to assert more regulatory oversight over food production. Not unlike the British perception that tuberculosis was primarily a problem associated with unhygienic living conditions, Americans associated trichinosis not only with undercooked pork but with feeding pigs waste products. The latter problem would be largely eliminated after hogs were increasingly raised in confinement after World War II. Meat-borne bacteria, such as *salmonella*, discovered in 1885, would become public concerns during the late twentieth century in the context of increasing health concerns about meat.[11]

After close to a decade of debate, economic not health concerns compelled the US Congress to approve two measures to address the trichinosis problem. A potential estimated loss of $260 million in pork exports over the period spurred this action. On August 30, 1890, Congress passed an act requiring microscopic examination of pork products for exports. Then on March 3, 1891, an additional act requiring inspection of slaughtered hog carcasses passed. The US government also threatened Germany with higher duties against its sugar exports before Germany rescinded its ban against American pork with the Saratoga Agreement on September 3, 1891. Other European countries lifted their bans as well. Interestingly, however, because of the German states' autonomy in regard to food inspections, local microscopic inspections of American pork sold in Germany continued. An Imperial Slaughter and Meat Inspection Act passed on June 3, 1900, prohibited importation of sausages, canned meat, and meat with preservatives. Then, in 1903, Germany prohibited the importation of live cattle and fresh beef from the United States. Starting in 1910, pigs could be imported from the United States, but only after a four-week-long quarantine, and were then subject to a high tariff. Surprisingly, passage of the US Meat Inspection Act of 1906 that focused on American domestic meat production practices postdated concerns about the safety of American meat exports.[12]

Outside the West, other national governments developed similar regulatory authorities during the late nineteenth and early twentieth centuries. In Japan, the notion of sanitary meat products remained complicated for decades because of the association of animal flesh and blood with Japanese prejudices about the "unclean" elements of the *Buraku* population. Nevertheless, the Meiji government passed the Slaughterhouse Act of 1906 aimed at both better managing the rapid proliferation of slaughterhouses through government licensing and ensuring safer access to sanitary meat, particularly to control the incidence of *cystic hydatidosis* or *echinococcosis*, caused by eating meat infected with parasitic tapeworms. Especially due to infected cattle imported from China, where *cystic hydatidosis* was endemic, as well as to cattle raised in Japan's southwestern prefectures, the Slaughterhouse Act of 1906 helped to dramatically decrease the incidence of the disease. Between 1881 and 1906, about forty-five cases were reported. From 1920 through the end of World War II, there were only two known cases.[13]

Of course, in Japan, China, and most of East Asia before the mid-twentieth century, regulation of meat for the purpose of promoting better sanitation was a minor issue due to meat's minimal dietary role. One indication of this fact in Japan was that while the incidence of pulmonary tuberculosis increased with the rise of urbanization, industrialization, and contact with the West after the mid-nineteenth century, bovine tuberculosis, with which Britain and much of the meat-eating West struggled, was almost completely unknown in Japan before World War II. Vaclav Smil and Kobayashi Kazuhiko estimated that in 1900, per capita annual meat consumption in Japan was just about one and one-half pounds. Fifty years later, it had risen only to slightly over three pounds.[14] In the first decade of the twentieth century, Russell Chittenden quoted Ōshima Kintaro:

> The rural population of the interior depends very largely or entirely upon a vegetable diet. Fish is eaten perhaps once or twice a month, and meat once or twice a year, if at all. The poorer working classes in the cities also use very little animal food. But the poorer classes in the city and the peasantry of the rural districts comprise nearly 75 per cent of the total population, and it is therefore safe to assume that this proportion lives chiefly, or wholly, upon vegetable diet. And this, it may be observed, means vegetarianism literally.[15]

Nevertheless, Japan, like much of Western Europe, turned increasingly to public oversight of meat production for reasons that were partly due to increasing sanitation concerns.

## State Regulation of Meat Production

Throughout Western Europe over the course of the nineteenth century, public, typically municipal, ownership of slaughterhouses became increasingly common. After Paris opened its first public slaughterhouses in 1818, other cities followed in France, Belgium, Austria, Italy, Switzerland, and Germany. Completion in 1867 of La Villette, Paris's grand central abattoir, came in the midst of similar large-scale municipal initiatives in London, Berlin, and many other European cities. Although pressure to nationalize the entire meat industry was developing in England, a mix of private and public ownership was retained through the twentieth century, as was also the case in Mexico and Japan. Canada took a step in the direction of municipal ownership with the Toronto municipal slaughterhouse, opened in 1914; but Canada like the United States remained dominated by privately owned meatpacking firms.[16]

The impetus for public ownership of meat production stemmed from multiple initiatives and concerns. As urbanization and industrialization progressed during the nineteenth century, governments in much of Europe and elsewhere saw infrastructure industries—telegraphs, railroads, electricity, gas, water—as natural monopolies that lent themselves to close regulation and even government ownership. Although patterns of government involvement varied tremendously—with countries like Germany and Sweden pursuing municipal ownership more aggressively than England or France, for instance—virtually all modernizing nations agreed that some degree of government involvement in certain key industries, including meat production and consumption, was important. In nearly all cases, however, private companies started these industries before governments acquired them. By the early 1900s, state or municipal ownership of railways, electricity, or gas was greater than 50 percent in the United Kingdom, Denmark, Germany, Sweden, Italy, and Norway. In contrast, both France and Spain had almost no government ownership of these infrastructure industries. Robert Millward's analysis of European government ownership of infrastructure industries suggests that countries where interest guarantees and subsidies were common were usually the most likely to see government ownership and that these tools were products of pragmatic concerns with economic growth and social and political unification. Millward stresses that ideological arguments about municipal socialism—like those that were free-flowing in the United States in the Progressive Era—were not common until the twentieth century. Indeed, nationalization of many municipally owned industries did not occur in most nations until the twentieth century.[17]

The variation among countries in terms of private versus public meat production reflects not only underlying differences in economic practices but also differences in concerns most typically about problems associated with meat production's environmental and health effects. For instance, in Western Europe, reformers of various sorts saw the small, private slaughterhouses as horrors; they scandalized urban residents with their sights, sounds, smells, and disease risks. After centuries of generally tolerating the smells associated with livestock and meat making, Europeans in the nineteenth century began to voice concerns about slaughterhouse odors as the industrial revolution unfolded.[18] British physician John Hogg "decried the existence of slaughterhouses 'reeking with gore' in London"; and another British physician Hector Gavin asserted that manure manufacturing located in a London suburb "gave out the most horrid, offensive, and disgusting concentration of putrescent odours it has ever been my lot to fall victim of."[19] In terms of the latter, after the mid-nineteenth century, officials and educated citizens were well aware, for instance, as noted earlier, that tuberculosis could spread from animals to humans. In Germany, because of the popularity of eating uncooked or undercooked pork, trichinosis was a great concern. Although government inspection was typically advocated and attempted in many European countries, the proliferation of hundreds of private slaughterhouses made inspection of all of them very difficult. For instance, in 1873 Berlin had 1,113 butchers, who operated 780 private slaughterhouses. Growing urban populations that demanded increasing amounts of meat made the task of insuring food safety and public health increasingly difficult.[20]

Thus, during the late nineteenth and early twentieth centuries, municipal slaughterhouses became established generally for the purpose of controlling and regulating the health and environmental conditions in which meat was made. Municipal slaughterhouses became commonplace in Britain, France, and especially Germany. Over the last quarter of the nineteenth century in Britain, France, and Germany, reformers were partially successful in transforming a private and largely unregulated system of provisioning meat into a more systematic and safer process. Municipal slaughterhouses were built as well in Argentina, where the largest public slaughterhouse was at the Liniers market in Buenos Aires, and in much of Australia, including the large market in Sydney. In New Zealand, private companies typically slaughtered animals, but the national government controlled much of the export trade through the Meat Producers Board established in 1922. Australia also developed a meat export trade board in 1922, but its powers were more limited. While its burgeoning meat industry was based largely on private firms, Japan too attempted to institute a more systematic process of municipal government control and oversight of its slowly growing meat industry during the late nineteenth and early twentieth centuries. Although proposed, municipal slaughtering remained almost unknown in the United States and Canada. Throughout much of the West, farmers' cooperatives were involved in some parts of the meat marketing process but more often in terms of retailing rather than slaughtering.[21]

Between 1828 and 1899, five separate slaughterhouse movements, according to Ian MacLachlan, worked initially to move the Smithfield Market outside the city limits and then to end private slaughter in England altogether. Historically, Smithfield Market had been a site for not only animal slaughter but also fairs and animal-baitings, as well as human floggings and executions. London's Worshipful Company of Butchers, one of the city's oldest guilds with origins in the tenth century, opposed the effort to move the market. They wanted to modernize Smithfield but leave it in place. The reformers' first goal was accomplished in June 1855 with the opening of Islington Market two miles north of Smithfield. This goal was achieved in part because the power of the butchers' guild declined during the 1840s and 1850s. In 1856, London's common council passed an act removing guild membership as a requirement for the city's meat trade.[22]

The growth of municipal slaughterhouses in other European cities, especially Paris's La Villette in 1867, created greater support for public slaughterhouses in England. Many butchers, especially retail butchers, continued to argue that small shops created more competition and better prices. The Slaughterhouse Act of 1874 provided for more regulation of the slaughter process but also made concessions for retail butchering. Indeed, in that year, London still had an estimated 1,429 slaughterhouses, most of which were quite small and combined with domestic households. Nevertheless, growing concerns about bovine tuberculosis continued to galvanize support for greater public control. Also crucial over the last quarter of the nineteenth century was the expansion of livestock and meat imports in England. Wide ranging animal-disease regulations required foreign cattle to be processed at foreign animal wharves, primarily at the Deptford abattoir. Imports of chilled and frozen meat also undercut the need for London's multitude of slaughterhouses.[23]

Formation of the National Federation of Butchers and Meat Traders' Association in 1888 was perhaps the last major effort by England's retail butchers to maintain the old tradition of smaller private slaughterhouses. The association also attempted to lobby against greater imports of foreign meat. Although their efforts helped to forestall the movement toward public slaughter for some time, private slaughterhouses were finally banned in London in 1927. Not until 1966, however, was complete veterinary inspection achieved throughout London. Large municipal slaughterhouses had already been opened in many Great Britain cities outside London, though, including Edinburgh in 1852, Manchester in 1872, Leeds in 1886, and Birmingham in 1895.[24]

France had been at the forefront of the public slaughterhouse movement when Napoleon I created the first such establishments in Europe. When La Villette opened in 1867, most Europeans considered it the greatest slaughterhouse and meat marketplace on the continent. However, according to Kyri Claflin, La Villette "was a municipal slaughterhouse that was hardly more than an agglomeration of *tueries particulières*, or private butcher stalls."[25] In contrast to England's case, where retail butchers generally opposed the public slaughterhouse movement and attempted to maintain traditional practices in private slaughterhouses, French authorities created public slaughterhouses that protected traditional butchering practices by giving them space and practical control. Between the 1880s and World War I, however, agrarian and social reformers challenged this system of artisanal *abattage* (animal slaughter) that focused on high-quality fresh meat items, instead of greater quantities of lower-priced, refrigerated, and more hygienic items to meet urban mass consumption demands. The initial centralization of the meat trade at La Villette created a rational system for provisioning the meat that French consumers demanded in the mid-nineteenth century; but it was no longer adequate by the turn of the twentieth century. Greater regional dispersion of public slaughterhouses with refrigerated storage facilities did not develop in France until after the World War I era.[26]

Germany developed its municipally-owned slaughterhouses at about the same pace as France. Frankfurt and Munich opened facilities in the 1860s, as did Hamburg in 1872. Berlin's public slaughterhouse did not open until 1881. In the late eighteenth century, Berlin actually had three municipal slaughterhouses, but two of the three were closed by 1810 due to obsolescence. When authorities abolished traditional guild control of the trade in November 1810, butchering dispersed throughout the city and went largely unregulated. The discovery of pork *trichinae* in the 1860s, however, gave new impetus to greater regulation. Dorothee Brantz stresses that the public health threat from the disease led scientific experts to agree "that food safety was a matter of public welfare that had to be addressed by municipal and state agencies."[27] Most experts felt that, as long as butchers operated in their own shops and private slaughterhouses, public welfare was in danger. On March 18, 1868, a new slaughterhouse law passed that supported the construction of public facilities for meat production. Nevertheless, the law did not require that such construction occur, and the number of private slaughterhouses actually increased in the 1870s.[28]

Berlin's very rapid population growth spurred reform not only in terms of slaughterhouse practices but along many other environmental fronts. From 1820 to 1870, Berlin's population grew fourfold and reached

826,937 by 1871. In that year, Berlin became the capital of the new nation of Germany, and the Berlin city council supported construction of a new public slaughterhouse in 1876. Construction began in July 1880, and the facility opened in early 1881. A series of laws passed in 1881 and 1882 required all the city's meat to be produced there. The number of butchers working in the new facility, Central-Viehhof, grew from 67 in 1881 to over 1,000 by 1900. Another 750 inspectors worked there at the beginning of the twentieth century. By 1890, Germany may have been the world's center of the public slaughterhouse movement with ninety-four total facilities.[29]

Given the important role of municipal ownership of infrastructure industries in some cities in the United States, it is somewhat surprising that public ownership of slaughterhouses did not exist in any significant way. In 1906, for instance, George Stewart Brown reported that eight hundred electric lighting firms were municipally owned. Certainly concerns about animal diseases and food safety were just as important in America's meat production and consumption history as elsewhere. Nevertheless, the regulatory emphasis was almost always narrower and generally confined to inspection. Because of city health authorities' concerns, the Union Stockyards in Chicago seem to have been the first major American stockyards to adopt an inspection system for animals moving from the railroads into the stockyards themselves. The meat inspection acts of 1890–1891 gave the BAI authority to carry out inspections at the major slaughterhouses involved in the export trade. These acts also authorized inspections of meat intended for domestic consumption but provided no funds for this. The Meat Inspection Act of 1906 increased funding for inspections for all cattle, hog, sheep, and goat products for both export and domestic consumption. In 1912, the Meat Inspection Division became a separate branch of the BAI.[30]

In Japan, promulgation of the Slaughterhouse Law of 1906 was aimed at both better managing the rapid proliferation of slaughterhouses and ensuring safer access to sanitary meat. Its provisions required licensing of all slaughterhouses. The law also was concerned with more effectively shifting meat production from private to public facilities throughout the country. The law stipulated that municipalities had to pay the costs associated with slaughterhouse construction and that they pay the costs associated with maintaining slaughter facilities and production. At the same time, municipal governments benefited from revenues associated with meat sales. In Ōsaka, for instance, after passage of the act, a new public slaughterhouse, the Kizu River slaughterhouse, opened in 1910. The privately owned Ōsaka Slaughterhouse Company, located just across the Kizu River from the new public facility, closed that same year. The new public slaughterhouse was a German-style, red brick plant in which German slaughtering techniques were reportedly implemented. From 1917 to 1924, the public slaughterhouse processed an average of twelve thousand to fifteen thousand cattle per year. Fees garnered from the public slaughterhouse were enough to provide about one-half of the village's annual budget. *Buraku* and other lower class workers from surrounding neighborhoods were hired as plant workers. In 1925, the city of Ōsaka took over the Imamura village slaughterhouse in the Nishinari ward of the city. Both plants operated under public ownership until consolidated into a new single facility in 1939. Despite these developments, the disparity in availability of meat in the United States versus Japan is revealing. For instance, in 1921, Chicago's privately owned packinghouses slaughtered 2,236,776 cattle compared to 13,099 in Ōsaka's publically owned Kizu River slaughterhouse. A few years before this, at the end of the Meiji era, two of the major public slaughterhouses in Tokyo slaughtered about thirty thousand cattle per year.[31]

In Western Europe, North America, and Japan, the state was instrumental in promoting greater meat consumption—especially through its efforts to control animal diseases, both epizootics that threatened livestock populations as well as zoonotic diseases that could be transmitted from animals to humans—and regulation of the slaughtering and meat production process. The primary emphasis of states during the nineteenth to mid-twentieth century in both regards was creation of greater public acceptance of meat as safe and wholesome to buy and eat. State intervention would continue to be crucial in reassuring the public about meat's safety long into the twentieth century. State regulations regarding creation of better working conditions for those making meat were generally a secondary concern but, as will be addressed in chapter 7, were crucial in helping to solidify greater meat production and consumption throughout the West and East after World War II. Only over the past generation has the state's role in promoting meat consumption taken a back seat to corporate efforts.

# 5

## Meat Culture Transformations and New Animal Products

By the early twentieth century, scientific findings and state intervention supported the expansion of meat in people's diets. Dietary advice converged on the idea that meat, particularly the protein in it, was necessary for good health. State policies increasingly promoted the security of wholesome meat products that consumers at all income levels would purchase. Meat producers wanted consumers to eat more meat, and they worked hard to develop meat products that consumers would want to eat. Traditionally, aside from meat choices influenced by specific ethical, religious, or cultural taboos, people ate whatever meat was available. But as the twentieth century advanced, modern tastes around the globe tended to both narrow primarily on fresh muscle meat products and converge on similar types of animals. Around the globe through the mid-twentieth century, beef was increasingly the premier meat choice. Poultry would become much more important during the late twentieth and early twenty-first centuries (as addressed in part III). The narrowing and convergence of global meat preferences before World War II spurred trade and meat's increasing dietary prominence after the war.

As more meat was produced and consumed, meat producers also focused increasing attention on the development and marketing of animal products other than meat, often known as offal or, as described here, by-products. From the late nineteenth through the mid-twentieth century, edible by-products, such as lard, became relatively less important among items that meat companies marketed while inedible by-products assumed greater significance. While traditionally valuable inedible by-products such as skins or hides and fats remained important in the twentieth century, a much greater array of inedible animal by-products emerged as significant among the industrializing and growing populations of the West and East Asia.

## Traditional Meat and By-Product Choices

Partly because most people, aside from the wealthy, ate very little meat, Westerners in the medieval and early modern era ate just about whatever animal flesh they could obtain. Beyond beef, mutton, pork, and poultry, which historian Ann Hagen explains was a "prestige food" and rarely eaten, people ate meat from an enormous variety of wild animals and birds.[1] Boars, hares, and bucks, especially, were widely consumed, as were fresh fish and molluscs when they could be obtained. As noted, consumption of wild game was important in Japan too. Throughout much of Europe after the eighth century AD, horseflesh became a taboo food. Horseflesh consumption, a practice known as hippophagy, was not widespread in Western history, as it was among peoples of the Eurasian Steppe, especially the Mongolians. Pope Gregory III censured the practice in AD 732 as part of the Catholic Church's attempt to convert Germans by ending what it considered pagan ritual practices. While the Church sometimes allowed the practice during times of famine, most Western Europeans avoided hippophagy primarily out of concerns about wholesomeness as well as increasing sentimentalism about the status of horses.[2]

The parts of animals and birds eaten in the premodern eras of the West and of East Asia were as varied as the types of animals and birds consumed. Joan Thirsk notes that, during the sixteenth and seventeenth centuries in England, the English "relished kidneys, livers, cheeks, tripe, noses, bone marrow, as well as the muscle meat of beef cattle, sheep and pigs. Crisply roasted sheep's heads were a special treat."[3] Calves' and pigs' brains were another delicacy. In addition to being the world's most important pork culture through the modern era, the Chinese were arguably the world's leading consumers of edible pork by-products. Dishes made from pig's brains, ears, tongue, lung, tripe, large and small intestines, kidney, heart, pancreas, bone marrow, womb, bladder, testicle, skin, and blood were all popular. The Casserole House Restaurant in Beijing, opened in 1741, was famous for its pork casserole that consisted in part of large intestine. Pigs' blood was cooked with some salt and cut into cakes. Lard, made from rendered pig's fat, was another traditional edible by-product in the West and in China. Although the Romans did not typically use lard in their cooking, once Germanic peoples inundated much of the Roman Empire, lard became more important and increasingly associated with preparations of meat over the course of the medieval and early modern eras.[4]

Before industrialization, the most valuable inedible animal by-products consisted primarily of hides, skins, and fats, all of which had had immense value for a variety of human needs for thousands of years. Decline of the North American fur and hide trade during the nineteenth century may have contributed to greater demand for livestock hides. Preparation of leather from cattle hides was practiced as a craft industry throughout Europe and North America until the 1840s, when some elements of manufacturing were gradually introduced. Virtually all settlements of some size in the British colonies of North America had a tannery along with sawmills and gristmills. Indeed, leather was always in great demand. Shortages occurred only when skilled leather-makers were lacking, since the hemlock and oak bark needed for tanning and cattle hides were abundant. In the early 1800s, raw hides from Argentina's cattle supplemented the tanning industry's supply in the United States and Europe. Japan's *Buraku* leather-makers, despite their segregated status and the discriminatory practices against them, received special protection during the fifteenth and sixteenth centuries because of their importance in producing armor for the samurai who were embroiled in the constant warfare of this period.[5]

Likewise, wool was vital for clothing in Europe. Especially during the medieval period, sheep's wool was more important than its meat. Historian Ephraim Lipson asserts that woolen manufacturing was England's most important industry from the twelfth to the nineteenth century. For most of this period, domestic sheep provided the wool for the industry, though Negretti and Merino breeds from Spain were imported by the nineteenth century along with foreign wool. Australian wool in particular became important to England's woolen manufacturing during the nineteenth century. In the latter part of the nineteenth century, the United Kingdom's imports of wool exploded. Between 1831 and 1875, imports increased tenfold, from just over 31 million pounds to over 360 million pounds. Nevertheless, domestic sheep still provided about two-thirds of the country's wool. Compared to the medieval period when sheep were valued primarily for their wool, sheep raised in the British Isles by the 1870s were valued mostly for their meat.[6]

Rendering the remains of dead animals provided the other traditional by-products, such as fats, oils, and bone material. Once muscle meat and skins were removed, the remaining carcass would be boiled down; and the resulting tallow, in the case of cattle and sheep in particular, would then be sold for other manufacturing purposes such as candle and soap production. The boiling-down process produced odors that resulted in creative efforts to confine and contain the activity, such as in the case of Busby's "stink boat"—a riverboat that manufactured tallow while docked on the Mississippi River in St. Louis beginning in 1871. Boiling down was frequently important for the purpose of salvaging some monetary value for animals condemned due to diseases or as a mechanism for raising prices if livestock numbers were too high. The latter situation occurred between 1843 and 1851 in Australia when wool prices fell during the country's gold rush era. In that period, thirty thousand tons of tallow were exported to Britain. Of course, whales were the quintessential by-product mammal of the nineteenth century: the meat had very little demand compared to the oil, bone, and baleen produced.[7]

Before widespread use of chemical fertilizers, animal manure was also invaluable for fertilizer. Unlike hides, wool, fats, and oils, manure was not typically marketed by farmers; instead, it was a necessary part of farm life that accompanied animal husbandry. Earl B. Shaw noted that for traditional Chinese peasants "the hog is probably considered a machine or converter into which he puts so much waste or raw material, receiving in turn manure which is indispensable to the realization of his crops."[8] Because of farming's expansion, the United States, Britain, and much of Europe began in 1841 importing Peruvian guano deposited mostly from two species of birds—the gray pelican and the white-headed gannet—for use as agricultural fertilizer. Unusually rich in ammonia, guano was particularly important as a fertilizer through the 1870s. In 1875, guano made up 1.2 percent of all animal by-product imports in Great Britain and Ireland, which was twice the value of other imported animal manures.[9]

## Increasing Beef Consumption

From the 1880s to 1930s, meat consumption in the United Kingdom grew steadily until the 1920s when the rate of growth began to slow. While British consumer demand for beef was steady in the aggregate over this fifty-year period, evidence suggests that lower and middle income consumers ate relatively more beef while the more affluent ate relatively less. Marketing of Liebig's extract as well as importation of less expensive foreign beef had helped to increase poorer consumers' taste for beef in the United Kingdom and throughout Europe. Benjamin Seebohm Rowntree and May Kendall's examination of English rural laborers, published in 1913, included detailed weekly menus of various families from a variety of locations across the country. In that year, on average, about one-quarter to one-third of the average English family's food budget was spent on meat. Rowntree and Kendall's work suggests that even the rural poor in England already had ample access to meat. A sample of seven of the families from seven different parts of England showed six of them had meat for at least one meal every day of the week. Four of the families specified types of meat eaten. Roast beef or roast pork was a Sunday dinner for each of the four. Monday's dinner consisted of cold beef or pork. Tuesday's dinner included fried pork or warmed beef. Wednesday's dinner options were stewed or hashed meat or suet pudding. Thursday's dinners contained meat or suet pudding or beef liver. Friday's dinner included fish in one case, no meat in two others, and leftover liver in one. Only one of the four claimed to eat any meat on Saturday, and in that case it was fish and chips, which was a growing convenience food before World War I.[10]

About twenty years after the 1913 survey, Rowntree conducted another survey of meat consumption among a range of families with different incomes in York. By the 1930s, while a fairly strong correlation existed between family income and amount of meat eaten, the most important finding was that, as income increased, families ate a greater variety of meat. The poorest family, for instance, ate 4.5 pounds of meat per week with roast beef comprising two pounds of that amount. The next higher income bracket family ate ten pounds of meat per week with roast beef making up 4.5 pounds of that amount. At the highest income bracket, the amount of meat consumed per family was actually slightly less (8.4 pounds) with the meat choices including mutton, rabbit, ham, bacon, sausage, tripe, liver, and corned beef. Based on this survey, it may be possible to surmise that wealthy English families were already moving toward less overall meat consumption, more meat variety, and less interest in beef as a prestige food, all of which would be characteristics of consumers' eating habits in developed nations after World War II.[11]

Australia and New Zealand's beef-eating inheritance from Britain is evident in its meat consumption statistics during the early twentieth century. Meat consumption continued to be much higher in both countries than in Britain. The average Australian in 1924 ate 227 pounds of meat, which was almost double the British average. Of that, nearly two-thirds was beef—a greater proportion than in Britain, where it was about half. Almost a third of the Australian meat consumption consisted of mutton and lamb, which was much greater than the British proportion of 17 percent. Pork consumption in Australia at 5 percent was negligible compared to Britain's. New Zealanders, averaging 307 pounds, ate even more meat than Australians. Beef made up 55 percent of this total, followed by mutton and lamb at 39 percent, and pork at just 9 percent of the total.[12]

Already significant beef eaters, urban consumers in Argentina and Uruguay became greater fresh beef eaters during the first third of the twentieth century. Until 1904, cattle slaughtered at *saladeros* exceeded that at *frigoríficos*, which reflected in part the regional population's reliance on beef jerky and other processed beef. At that time, Argentina's fresh beef marketed in England was cheaper than that sold in Buenos Aires. During the World War I era, only about a quarter of domestic meat supplies came from the export-oriented *frigoríficos*. While combined frozen and chilled beef exports peaked in the mid-1920s, cattle slaughtered for domestic consumption during the 1920s and 1930s climbed steadily. For instance, during the 1920s, Liebig's domestic share of its sales jumped from 54 to 68 percent. By 1935, almost three-quarters of total cattle slaughtered in Argentina were consumed there. The estimate for per capita meat consumption in Buenos Aires in 1925 was 310 pounds, with 85 percent of this total from beef. The main national meat dishes consisted of *asado*, beef roasted in a coal fire; *puchero*, meat boiled with vegetables; and *locro*, a stew made from fresh or salted meat boiled with corn and other vegetables. *Carne con cuero*, beef roasted with the hide on, was a popular festival food. Meat consumption in Uruguay's capital of Montevideo also increased rapidly during the 1920s.[13]

Germans, Danes, Americans, and Canadians were the dominant pork consumers in the West during the nineteenth century. Although a preference for pork would continue in each country until the World War II era, beef was rapidly making inroads on this dominance. Evidence for this can be found in cookbooks, including Henriette Davidis's *Praktisches Kochbuch* (*Practical Cook Book*), arguably the most important German cookbook of the nineteenth century. Published in 1844, the first American edition appeared in German in Milwaukee in 1879, and it went through at least fifteen reprints, revised editions, and translations by 1911. Its first English-language translation came out in 1897. The number of recipes devoted to pork in the third English-language edition, published in 1904, is actually much fewer than those devoted to beef and veal, and even fowl. Hams and baked pork dishes are numerous, as are the variety of sausage preparations. Pork croquettes in South German Style—*Sueddeutsche Schnitzchen*—is the one of the few pork recipes referred to in German. One reason for the great number of beef and veal recipes is the large number devoted to variety meats, including several devoted to tongue, kidney, sweetbreads, liver, calf's brain, and calf's head, reflecting traditional uses of beef by-products among the masses. The number of meat soup and meat pie recipes equaled those devoted to other meat recipes. One important use of pork was in larding beef roasts. Long strips of pork fatback were sown to the beef roasts with a larding needle to make them juicier and more appealing.[14]

Americans' pork-eating proclivity was partly a product of its German immigrant population, but this characteristic also developed due to livestock practices and regional tastes. During the eighteenth and early nineteenth century, pork was favored in both the North and the South and, especially, among rural populations.[15] Whereas swine had traditionally been allowed to forage on beechnuts and acorns in mast forests in German and other European traditions, pig foraging was also encouraged in urban areas, including even in New York City, where, according to Catherine McNeur, "[h]ogs subsisted, for the most part, on the garbage that New Yorkers placed on the sidewalks and streets outside their homes and businesses."[16] Many people on the city's outskirts kept piggeries—property with penned pigs—until the 1840s. Not until 1849 did the city pass an ordinance outlawing free-roaming pigs.[17]

Pork was especially important in the Southerner's diet during the nineteenth and early twentieth centuries. Americans as well as many foreign visitors considered a Virginia ham the country's most outstanding meat item.[18] Sam Bowers Hilliard notes that the "most obvious characteristic of southern food habits was the strong dependence upon corn and pork which persisted throughout the nineteenth century and, indeed, well into the twentieth."[19] Pork was equally popular among whites and blacks in the South. Slaves subsisted primarily on corn and pork, and the South was the United States' leading corn- and hog-raising region from 1840 to 1860.[20]

Nevertheless, some early-twentieth-century Southerners promoted beef as a more desirable meat. Booker T. Washington was one of many American leaders at this time who advocated the importance of beef in human nutrition and as a foundation for civilization. As a result of Washington's personal insistence, the Tuskegee Institute served beef at one or more of the three daily meals for its African-American students during the early twentieth century. Washington hired Alice J. Kaine, a white home economist from Wisconsin, as the institute's nutritionist. She felt that beef was not only a nutritious food but also an "instrument of civilization."[21] As a consequence, the students who attended Tuskegee during this period ate beef in amounts virtually unknown among southern blacks at the time. Additionally, consumption of beef rather than pork, which was still associated with slavery, was one way that "deceitful Pork chops must be dethroned in the South," in the words of W. E. B. Du Bois in 1918.[22] During the first half of the twentieth century, throughout the United States, growing numbers of more affluent urban consumers led the way in choosing beef over pork; but it was not until after World War II, when widespread access to home refrigeration became available, that beef would surpass pork as the preferred American meat choice.[23]

North American hog slaughter was also vital in producing lard, which Kelly J. Sisson Lessens has described as the "defining element" in American cookery during the 1870s and 1880s.[24] North Americans as well as Western Europeans and East Asians ate lard for its flavor and the flavor it imparted in a variety of dishes. Farmers in the United States had raised lard-producing hogs since the country's founding. During the mid- to late-nineteenth century, most large US meatpacking firms sold hog fats to rendering companies, which produced lard. The Chicago-based N. K. Fairbank Company was the predominant lard-producing firm in North America. In 1917, Fairbank had five plants in the United States and Canada—Chicago, St. Louis, Gretna (New Orleans), New York, and Montreal—that produced lard and other edible fats, as well as soap products, distilled fatty acids, glycerine, and hydrogenated oils. By the 1870s, however, the major meatpacking firms also began to produce their own lard products. Leaf lard—the finest quality lard, which came from the hog's visceral fat around the kidneys and loins—was the most valuable grade of hog fat. Steam rendering converted fat from the remaining parts of the hog into "whole-hog" or "prime steam" lard that was edible. Lower qualities of lard were used to make lard oil, used in both burning oils and in making oleomargarine. The lowest quality lard was used in soap making or in grease oil. By the 1920s, the higher qualities of lard were increasingly clarified and bleached white—without taste and odor—to make a product that had a long shelf-life. Average per capita lard consumption among Americans rose steadily from 11.8 pounds in 1907 to 14.3 pounds in 1928.[25]

Lard also had valuable export value for American meatpackers during the late nineteenth and early twentieth centuries.[26] The meat industry's by-product expert, Rudolf Clemen, noted in 1927 that "[f]or more than a century and a quarter, the United States has been the principal country of surplus lard production, and the chief contributor to international trade in this commodity." Moreover, "exports of lard form the largest single export item derived from farm animals, and American lard is a dominant factor in the world's market."[27] Lard rapidly increased in value as a North American export and British import in the 1870s. The United Kingdom imported over 60 million pounds of lard from the United States in 1875. By 1900, this figure topped 200 million pounds. In part, the United Kingdom imported increasing amounts of US lard because their pork industry, like Denmark's, had transitioned to production of lean "bacon" hogs, which were poor producers of lard. After trade barriers and World War I impeded exports, US lard sales picked up again following the war and through the early 1930s.[28]

After production and consumption increases during the early twentieth century, several factors converged during the 1930s to decrease lard consumption. The US meat industry and swine breeders were beginning to focus on ways to improve pork meat sales by transforming hogs from lard-type to meat-type animals, partly to be able to meet European consumer demand. Although this transformation was not completed until after World War II, the industry's emphasis on lard sales declined as a result of this transition. Western consumer tastes shifted away from lard after mid-century to vegetable oils, in part due to growing concerns about saturated fats (as described in chapter 9). Already by 1937, American consumption of lard had declined to 10.5 pounds from its peak in the late 1920s.[29]

American consumers' taste for pork as well as lard began to decline during the 1930s. Although pork was also canned, consumers' preference for beef was evident in canned beef sales. Many major firms in Europe and North America offered canned meat during the 1870s. Canned and potted beef replaced sales of salted beef throughout the Western world in the 1880s. By 1910, 40 percent of the 795 federally inspected US meat-packing plants produced canned-meat products. World War I stepped up demands for canned meat, including products like corned beef hash. But globally the peak of canned meat and other food products was not reached until the 1950s. Canned meat was not only often cheaper and more convenient but by mid-century was a much improved product, in terms of both meat contents and packaging. The peak in canned-meat consumption in the United States came in 1956, when Americans consumed an average of eleven pounds per capita. At that time, outside the United States, consumers of the largest quantities of canned food, including nonmeat products, lived in the United Kingdom, France, Germany, Italy, Japan, India, and the Netherlands. After mid-century, though, refrigerated and frozen products would begin to overtake canned products across much of the globe.[30]

While still quite limited, Japanese dietary choices included more meat, especially beef, during the first half of the twentieth century. Between 1900 and 1940, Japan's estimated per capita meat consumption grew from two to four pounds per year. Historian Masuoka Jitsuichi surveyed typical Japanese food preferences in a report on Japanese immigrants in Hawaii. He noted that in 1900, the typical Japanese immigrant to Hawaii ate steamed rice, miso soup, and thinly sliced, pickled daikon (radishes) flavored with *shōyu* (soy) sauce for breakfast; cold rice, kidney beans with salt, and a potato-based stew for lunch; and steamed rice, daikon, and beef stew or dried fish for dinner. Although in Japan at the same time, daily consumption of meat would have been almost unknown, the consumption of beef stew (*gyūnabe*) was increasingly common and was, perhaps, the most popular way beef was eaten. Another type of stew, *sukiyaki*, consisting of various vegetables as well as thin-sliced beef, gained popularity by the beginning of the twentieth century. Food historian Ishige Naomichi estimated that in the 1930s the average Japanese might have a *sukiyaki* dinner once per month. As in China—as well as in many Western nations by the early twentieth century—soups, stews, and meat-flavored dishes were a way to stretch the available meat supplies among people who wanted meat but were not able to afford much or were not accustomed to eating much meat. Before World War II, Japanese tended to prefer beef from domestically raised *wagyū* cattle. However, even to meet the light beef demand of the population, much of the beef came from cattle imported from Korea and China. *Tonkatsu*, or pork cutlet, often served with curried rice, started to become popular during the interwar period in urban locations. But both pork and chicken, which would become especially important areas of Japanese agricultural development after World War II, were not eaten much before World War II.[31]

Although not directly related to rising beef consumption, use of horsemeat in parts of Europe, especially France, during this period helped to meet rising overall meat demand, especially among lower-income consumers. In 1866, French decree legalized slaughter of horses in designated abattoirs and allowed sale of horsemeat. Emile Decroix, chief veterinarian of the French Army and later the director of the country's leading veterinary school, Maisons-Alfort, was particularly important in promoting *chevaline* consumption. Horsemeat had few of the disease problems associated with beef, such as bovine tuberculosis, which had also contributed to fluctuating problems with beef supplies. Also contributing to the change in outlook was the rapid acceptance of *chevaline* consumption in several other European countries during the first half of the nineteenth century. While Iceland and Russia had never banned its sale, Denmark approved it in 1807; various German states had done so in the 1840s, Austria and Belgium in 1847, Switzerland and Prussia in 1853, and Norway and Sweden in 1855. Holland, Portugal, Turkey, Greece, Spain, Italy, and Great Britain still prohibited it, but in 1868 A. S. Bicknell reported that two hundred thousand pounds of *chevaline* were consumed in London each week. The fact that horsemeat consumption in Europe was gaining acceptance during the same period that Europeans in general were rapidly consuming more meat in general is not likely coincidental. Indeed, Bicknell advocated the sale of horsemeat primarily so that the poor could eat more meat. Consumption increased, especially during the interwar period. In the 1930s, the Netherlands imported nearly three million pounds of horsemeat per year from the United States—nearly the entire amount produced in the United States.[32]

## Diversifying Inedible By-Product Marketing

Expansion of various animal by-product uses accompanied rising global livestock populations and meat consumption. Meat companies increasingly focused on the marketing potential of animal parts that had traditionally gone to waste or turned into tankage for fertilizer. In particular, Western inedible by-product production and marketing evolved and expanded from the 1890s through the mid-twentieth century. Rudolf Clemen's exhaustive study of the by-product component of the US meatpacking industry in the 1920s highlighted four types of inedible by-products that the packing companies' more systematic development had elevated in value: fats, oils, and greases; pharmaceuticals; canned dog food; and poultry feed. Packing companies realized that while the value of the meat in cattle, hogs, and sheep constituted anywhere from 81 percent for sheep to 97 percent for hogs of the animal's overall value, the remaining animals' by-products, if systematically used and marketed, could profitably add to the packers' overall sales. The new emphases were also driven partially by the decline in two traditional inedible by-products, hides and wool. During the early twentieth century, the United States faced considerable international competition in leather production. By the early 1920s, imports of foreign hides were increasingly reexported to Europe. The US leather market was also depressed due to use of cheaper leather substitutes. Use of rubber instead of leather in boot and shoe heels had become standard practice, and machinery belting was increasingly made from other fabrics and rubber. The US wool market was also rapidly shrinking. Peak wool production was reached in 1909, and in the 1920s only about 15 percent of pulled wool in the United States came from meatpacking plants. Synthetic products were also replacing many of the uses of hides and wool.[33]

To expand their by-product sales, the larger meatpacking companies hired chemists to help them analyze and develop new products for parts of animals that had formerly been rendered or wasted. Armour and Company was the leader in this transition to greater by-product production and marketing in the United States The Armour Institute in Chicago, founded in 1893, was the forerunner of the industrial research facilities that virtually all other major US meat companies developed during the twentieth century. Moreover, meatpackers developed ancillary industries or investments in several fields tied to their new by-product emphases, such as the cottonseed oil industry, lard compound manufacturing, butter and margarine manufacturing, canning, poultry, and pharmaceuticals.[34]

Inedible fats, oils, and greases had longstanding value as meat industry by-products. Inedible tallow was important in the making of candles, though this use declined during the late nineteenth century. American exports of tallow to the United Kingdom increased from about 42 million pounds in 1875 to 63 million pounds in 1900. Other inedible fats, oils, and greases were increasingly valuable after 1900, including lard oil, a by-product of prime steam lard production, which was used as high-grade burning oil, thread-cutting machine lubricant, and leather oil, as well as for its stearic acid, alongside paraffin in candle making. Soap manufacturers were the most significant buyers of high-grade inedible tallow. Traditionally a product of household production, by the mid-1920s the United States had about 270 soap factories, all of which relied on inedible animal fats, particularly glycerine, for one of the chief ingredients in their products. Neat's-foot oil, made with tallow obtained from cattle's feet, was an important lubricant for a variety of machinery.[35]

Among the foremost North American meatpackers, development of pharmaceutical products emerged largely as an outgrowth of their increasing involvement in the making of margarine and lard compounds. During the 1880s, all the leading American meatpackers hired chemists to investigate and develop medicinal uses of animal glands. Pharmaceutical laboratories in meatpacking plants became commonplace. In 1887, pepsin became the first pharmaceutical item extracted for its medicinal value by meatpackers. Laboratory staff sampled and analyzed animal products not only for possible commercial value but also as a part of the quality control operations of meat manufacturing. In 1889, French physiologist C.-E. Brown-Séquard injected extracts of animal testicles into an ailing seventy-two-year-old who then recovered. The English physician George Oliver is credited with the discovery of adrenalin from adrenal glands in 1893; and thyroxine from thyroid glands was discovered in 1895. These developments, combined with earlier industry efforts, inspired the major American meatpackers to produce proprietary lines of so-called organotherapies. Armour, for instance, produced organotherapy products under the brand name Glanoid.[36]

During the 1920s, US meatpacking companies' extraction of animal glands initiated a new era of therapeutic treatment in medicine. While many major meat firms continued to pursue their own lines of organotherapy products, they also provided extracted glands for scientific research and pharmaceutical company development. The first major alliance between meat firms and the pharmaceutical industry occurred with the discovery and production of insulin. Discovered by researchers at the University of Toronto in 1921, insulin was refined from both hog and cattle pancreases. Injected into diabetic patients, insulin revolutionized their treatment; instead of a fatal disease, diabetes became a manageable disease. Soon after its discovery, Eli Lilly and Company in Indianapolis collaborated with Chicago's major meatpackers to mass-produce it.[37] Researchers Tricia Close-Koenig and Ulrike Thoms characterized this insulin production alliance as "a major success for the relatively young American pharmaceutical industry, leading to a shift from Europe as the major place of pharmaceutical research and production to the USA."[38] Also in 1921, American researcher Herbert Evans discovered that pituitary glands could be used for making growth hormones. This led to a relationship with Armour to provide him with pituitary glands during the 1920s. Nevertheless, Armour continued to produce its own pituitary-based organotherapy products during this period. They sold a proprietary version of beef growth hormone until 1949.[39]

Rudolf Clemen highlighted the rapid expansion of pharmaceutical efforts by American meat companies. In 1923, he noted that meatpackers produced forty-eight pharmaceutical products for the medical profession and explained that the six most important were pepsin, made from hog stomachs to aid with dyspepsia; pancreatin, made from hog pancreases and used to peptonize food for infants and invalids (note that use for making insulin had just been developed); thyroid glands from sheep to treat people with thyroid conditions such as goiter and cretinism; epinephrine, made from sheep's suprarenals for use in treating asthma, hay fever, and hives as well as an astringent and heart stimulant; pituitary from calves for surgery and obstetrics; and rennin from calves' stomachs for both pharmaceutical purposes and cheese-making. The major US meat companies generally maintained laboratories at multiple meat production facilities. For instance, Swift and Company maintained laboratories at eight of their larger packing plants around the country. Just four years later in 1927, Clemen estimated that already at least one hundred pharmaceutical products were made in packinghouse laboratories. In that short time span, insulin, for instance, had become one of the most valuable animal by-products, though in its case animal pancreas was primarily supplied to pharmaceutical companies, particularly Eli Lilly, for manufacture into insulin.[40]

Through the mid-twentieth century, the list of pharmaceuticals made from animal glands continued to expand, including the hormone ACTH, made from hog pituitary glands and used in treating arthritis. Herbert Evans was also centrally involved in the development of ACTH starting in 1944 when his laboratory received federal government funding to isolate and purify its production. Once again Armour supplied Evans with pituitary glands for this research, which led to the successful development of the hormone for use in treating rheumatic fever and rheumatoid arthritis. At the same time, Armour's research labs developed their own version of ACTH, which it called ACTHA, starting in July 1948. Indeed, Armour opened a new laboratory complex in 1953 at a cost of $12 million that would also become its major pharmaceutical manufacturing center. Outside the United States, Ernst Laqueur, a German biochemist and researcher at the University of Amsterdam specializing in sex hormone research, reached an agreement with Saal van Zwanenberg, director of a Dutch slaughterhouse, to create a new company, Organon Limited Company for the Manufacture of Organ Preparations on a Scientific Basis. Laqueur was the company's director starting in 1923.[41]

The commercial dog food industry started in England in the 1860s with Spratt's Patent Limited. Its founder, American James Spratt, developed biscuits that were marketed primarily to sportsmen and dog fanciers. Most dog owners around the world through this period simply fed their animals table scraps. Spratt's Patent Limited established manufacturing facilities in the United States later in the nineteenth century. Its most famous product was the Meat "Fibrine" Dog Cake, which was broken up and moistened before being given to dogs. A few other firms offered biscuits for dogs by the end of the nineteenth century, but Spratt's was the dominant company, offering products to a limited consumer base. Starting in the 1910s, livestock feed manufacturers in the United States also added dry dog food to their product lines. The most well-known of these manufacturers was Ralston Purina Company, which began its dog food production in 1926. By the early 1930s, as the canned dog food business expanded, four horse-slaughtering companies in the United States canned their own product, and several other dog food companies purchased horsemeat.[42]

American meat companies launched canned-meat production for dog food during the 1930s. As historian Katherine Grier points out, the expansion of canned dog food sales during the Great Depression remains mysterious. Why would a relative luxury purchased for animals become popular when so many people lived in poverty? The most likely answer is that meat companies were all vigorously expanding their by-products during the interwar period and saw animal feeds made from tankage—the term for the residue left when all other useful meat materials have been rendered—as a new by-product line with sentimental appeal. Tankage as well as edible meat by-products, which were increasingly less desirable among American consumers, could be combined into canned dog food contents. John Morrell and Company introduced its Red Heart brand in 1932 and by 1934 spent considerable money on advertising its three varieties: beef-flavored Diet A, fish-flavored Diet B, and cheese-flavored Diet C. During the 1930s and 1940s, many studies found that the protein content of canned dog food was minimal, generally around 5 percent by volume, suggesting that a good deal of vegetable-based fillers were mixed with the meat. Yet by the early 1940s, an estimated 20 percent of Americans fed their dogs commercially canned products. World War II regulations prohibited use of cans for dog food, but once the war ended, canned dog food production and sales expanded again. By 1953, about three-quarters of Americans fed their dogs commercially canned products.[43]

American meat companies also started to pursue production of poultry feed from tankage and other meat processing during the 1920s. Rudolf Clemen noted that the poultry industry had leaped from negligible importance to considerable size during this decade. By 1926, the poultry industry equaled about 16 percent of the total value of the livestock industry. Although poultry's significance in the global diet is a post–World War II development, American meat companies saw potential in pursuit of poultry feed though not much yet in terms of poultry meat production. Before World War II, the major packers did not engage in slaughter or processing of either chicken or turkey. Nevertheless the period from the 1920s to the 1940s laid the groundwork for poultry's postwar explosion in production and consumption, as explained in chapter 6.[44]

Meatpackers' intensification of by-product production continued to evolve after World War II. Although some previously profitable by-products remained so, many others declined. Lard largely disappeared because of health concerns among Westerners, and synthetics steadily replaced drugs such as insulin following World War II. Plastics supplanted many of the uses for animal products as well. Leather and wool use in shoes and clothing declined steadily. Cellulose casings steadily superseded use of animal casings for sausage-making. Nevertheless, the changing market in both edible and inedible animal by-products after World War II helped to further cement the global growth of the meat industry and trade among meat-producing nations. While Americans largely lost interest in edible by-products, US meat firms found new markets in Western Europe, Latin America, and East Asia for edible pork and beef by-products. China increasingly imported North American leather, which helped it to augment its own production and make it one of the leading producers of leather goods by the end of the 1990s.[45]

# PART III

# Meat Becomes a Global Food

## Consumer Acceptance and Resistance after World War II

# 6

# The Political Economy of Meat after World War II

WIDESPREAD MEAT EATING BECAME A GLOBAL PHEnomenon after World War II. Starting in the mid-nineteenth century in the West, consumers at all income levels wanted to make meat a more significant part of their diets. This desire characterized consumption patterns among increasing numbers of people throughout nearly all parts of the globe after the mid-twentieth century. National governments' policies and corporations' creation of new markets transformed meat into a global consumer commodity through expanded and Western-dominated trade relationships and new developments in animal agriculture after World War II. A reordering of the global food regimes occurred immediately after the war, which facilitated the structural expansion of meat production and consumption outside the traditional meat-eating countries of the West. New trade agreements cemented provisions for moving grain—especially from North America to other parts of the world—that then could be used to feed livestock for developing nations' meat industries. In addition, many of the world's leading meat-producing countries increasingly traded meat products directly with more countries, especially in Asia. Because of these developments, worldwide consumption of red meat and poultry, based on the United Nations' Food and Agricultural Organization (FAO) calculations, nearly doubled from an average of almost forty-nine pounds per person in 1961 to ninety-one pounds per person by 2011.[1] (See tables 1 and 3 in appendix.)

Accompanying the new trade relationships was a major shift in meat consumption patterns that started in the West and then spread around the world. Poultry, especially chicken, became the preferred meat in much of the world by the end of the twentieth century. Chickens were transformed from animals that were consumed primarily on special occasions at high prices into animals that could be eaten regularly and cheaply. Tied to environmental and health developments that are explained in chapters 9 and 10, chicken toppled beef demand in part because it became cheaper and more convenient and was thought to be healthier. At the same time, beef increasingly lost prestige value among Western consumers due to its transformation into fast-food hamburgers and concerns about its environmental and health effects. Between 1961 and 2011, world average consumption of beef was flat, while poultry rose more than 500 percent. (See tables 1 and 3 in appendix.)

This chapter focuses special attention on the ramification of these developments in the phenomenal rise of meat consumption in Japan and China. In contrast to the United States, for instance, where meat consumption grew by about one-third, Japan's meat consumption increased tenfold and China's almost twentyfold between 1961 and 2011. (See tables 2 and 4 in appendix.) Cultural changes in beliefs about meat's nutritional value and ethical status, as well as state promotions, laid the groundwork for these changes. Consumption levels rose dramatically after the war due to greater demand that was met through trade, state promotions of meat production and consumption, and availability of new meat products.

## Livestock Feed Grain and Meat Trade during the Second Food Regime

Although the cultural disposition of much of the world's population—including East Asia—to want to consume more meat originated before World War II, the specific structural conditions that allowed for this increased consumption to occur, particularly in terms of world agricultural trade, developed during the second food regime from the late 1940s through the early 1970s and then continued to evolve during the corporate food regime that replaced it. The second food regime originated in the immediate aftermath of World War II with the rise of the United States to world economic dominance. As part of this development, the

United States was instrumental in establishing the General Agreement on Tariffs and Trade (GATT) that went into effect in 1948. GATT substituted relatively freer trade relationships for the protectionist environment that developed during the interwar period. US grain exports in particular became central to world agricultural trade restructuring. Not only did US wheat production make up for shortages affecting European consumers, but US corn and soybean production also helped to provide the necessary feed (or coarse) grains to build up livestock production in Europe and, later, in Asia. (See table 5 in appendix.) North America's role in providing feed grains to much of the rest of the world started during the mid-1930s and expanded through World War II. Elsewhere during the 1930s and 1940s, world feed grain production was stagnant or declining. Before World War II, developing countries provided 60 percent of world exports of feed grains compared to 20 percent by developed countries. Also before the war, developing countries imported only 2 percent of the world's feed grains. These trade relationships reversed after World War II, when Africa and Asia in particular became net importers of grains and meat. Those regions also lost export markets for tropical goods that had been crucial to their economic development earlier in the twentieth century. From 1948 to 1952, the United States' Marshall Plan in Europe accelerated grain trade—since 29 percent of the $13.5 billion in aid consisted of food, feed, and fertilizers—which then continued to increase after passage of Public Law 480, the Agricultural Trade Development and Assistance Act of 1954. This expanded US grain trade was normalized in 1966 with the Food for Peace program.[2]

By the early 1970s, the United States, and North America more generally, dominated world trade in feed grains. (See table 6 in appendix.) Roughly two-thirds of US corn production went to animal feed. Between 1955 and 1969, developing nations' share of global agricultural exports fell from 45 to 33 percent. By 1970, the major grain and meat exporters, including the United States, Canada, Australia, and New Zealand, accounted for the bulk of world agricultural exports. The United States alone made up about one-sixth of this trade, which consisted of nearly $6 billion in food and livestock feed in 1970. In the mid-1970s, the United States accounted for about two-fifths of the world's trade in wheat and rice, one-half of the trade in feed grains, and 90 percent of the trade in soybeans. The confluence of increasing availability of feed grains for livestock and growing consumer demand doubled world beef prices between 1963 and 1971.[3]

While US dominance of feed grain exports boosted many nations' livestock production, North and South America and, increasingly, Australia and New Zealand added to the world's meat supplies through exports of meat during the immediate postwar years. Europe experienced food shortages in general, including meat, through the early 1960s. For example, during the first half of the 1950s, British meat consumption was 30 percent less than it had been before the war. Meat rationing continued there until July 1954. International trade fluctuations significantly affected the United Kingdom, which remained the world's largest food importer in the immediate postwar years. In particular, Argentina's role in supplying the United Kingdom with meat, which had started to decline in the 1930s, continued to decline after World War II. Through the mid-1950s, Australia and New Zealand supplied the United Kingdom with the bulk of their meat production, particularly frozen mutton and beef. Starting in the late 1950s, both countries diversified their trade to include the United States, several European nations, and Japan, so that by 1961, Australia and New Zealand were the world's dominant meat exporters. Although North and South America combined were net exporters of meat, for much of the post–World War II era, the United States was actually a net importer of meat since domestic demand, particularly for cheaper ground beef, grew rapidly. Argentina was the largest exporter of overall cattle meat in 1961 in North and South America, while Mexico was the largest exporter of boneless beef in the same year.[4]

State support for agriculture was a cornerstone of the second food regime; and government policies made meat production a priority, especially in Europe and Japan. Government support for US agriculture, including trade policies, became ensconced in the 1930s with a variety of domestic programs as well as the Reciprocal Trade Agreement Act (RTAA), passed in 1934, which gave the executive branch and, specifically, the State Department, authority to bargain tariff rates. Starting in the 1950s many European countries' governments provided similar assistance that then became, according to Christina Davis, "entrenched" in Europe through establishment of the Common Agricultural Policy (CAP) in 1962.[5] Japan's Agricultural Basic Law of 1961 was also significant in giving livestock production much more support and protection. As a result of the GATT Kennedy Round of negotiations from 1962 to 1967, CAP targeted European self-sufficiency in food based in part on importing American soybeans to promote livestock production. Then, during the decade of the 1960s, CAP increased feed grain production among its member nations by just over one-third and more than doubled poultry meat production. While European Community and CAP support for feed grain caused tensions with the United States, especially given that feed grains constituted nearly one-half of the value of US agricultural exports in 1974, with serious repercussions in GATT negotiations in the 1960s and 1970s, CAP policies transformed the European Community from a net importer of meat in the 1960s to a net exporter by the 1980s. (See tables 6–8 in appendix.) It did so at a high cost to consumers in Europe. Nevertheless, the European public typically defended CAP policies in this period as necessary for the preservation of farm life.[6]

From the early 1970s through the GATT Uruguay Round (1986–1994) and the establishment of the World Trade Organization (WTO), which commenced in 1995, a corporate food regime dominated by transnational corporations (TNCs) increasingly replaced the state system of world trade that had characterized GATT and the second food regime. The TNCs needed states in terms of access to markets and capital but otherwise became increasingly powerful in determining world trade in feed grains and meat, as well as virtually all other agricultural commodities. As TNCs such as ConAgra, Cargill, Archer Daniels Midland (ADM), Continental Grain, and Central Soya transformed independent producer production around the globe into increasing contract and corporate production, they expanded the degree of vertical integration and agroindustrialization. As noted in the chapter's next section, this transition became especially characteristic of the poultry industry but made enormous strides in the pork industry and to a lesser extent with the beef industry. Mergers in the food industry, particularly during the 1980s, gave a relatively small number of corporations control over the production, processing, and exporting of many agricultural commodities. Although most of the major firms were US based, joint ventures with firms around the globe gave these TNCs a truly worldwide reach. By 1988, for instance, Cargill had 841 operations in forty-nine countries. During the same period, the twenty-five largest corporations controlled two-thirds of the global food processing activity.[7]

Transnational corporations' dominance of the world's feed grain, livestock, and meat trade had dramatic effects on meat trade relationships within the Americas. Starting in the 1960s and accelerating through the 1970s and 1980s, Latin and South American countries became home to increasing numbers of TNCs, which increased these nations' dependence on foreign corn, wheat, and soy for processed meat production, primarily for affluent countries' markets, at the expense of traditional staple food production. By 1981, American grain exports reached 118 million tons, or 56 percent of world trade, and approximately one-third of US farmland was used to produce primary grains for exports. Whereas Australia and New Zealand had been the major sources of cheaper boneless beef for the American hamburger market in the late 1950s, increasingly Latin and South American countries became the sources of these American imports of cheap beef products, especially for the fast-food market. This transition contributed to the slash-and-burn agriculture that became associated with the destruction of rainforest in the Brazilian Amazon. Mexico's traditional role in providing feeder cattle for finishing in the United States reemerged. At the same time, Mexico became a major importer of American edible by-product meats, due to declines in US domestic consumption of these types of meat. In 1987, most of the American and Canadian edible beef and, especially, pork by-products went to Mexico. Between 1972 and 1979, Mexico increased its pork production by 65 percent, poultry by 35 percent, and beef by 32 percent. To do so, it increased its sorghum production for animal feed substantially. The North American Free Trade Act followed by the US Farm Bill of 1996 increased the amount of US animal feed and meat products sold in Mexico. Smithfield Foods allied itself with Granjas Carroll to increase Mexican pork production, while it also sold more American-slaughtered pork there. Between 1994 and 2012, pork imports cost an estimated 120,000 Mexican farm jobs. Over the past generation, peasant agricultural economies in Mexico and much of Latin and South America increasingly disintegrated in the face of this animal feed and meat trade, leading to the movement of peasants to urban areas to seek employment in industry, as well as increasing the numbers hoping to find new economic opportunities in the United States.[8]

The corporate food regime also transformed European and Oceania meat trade. During the 1970s and 1980s, Australia and New Zealand lost much of their market in the United Kingdom, partly due to the UK's new trade relationships with Europe. As a consequence, Australia and New Zealand, especially, pursued new markets for their sheep meat in Russia and the Pacific Islands. For example, fatty "flap" sheep meat increasingly became a staple in the diets of Papua New Guineans. Indeed, as CAP helped to make Europe increasingly self-sufficient in meat, the role of Argentina and Uruguay in supplying fresh beef to Europe declined, while cheaper boneless beef production rose in those two countries and Brazil's and Latin America's share in European markets grew. In 1978, Argentina's share of world meat exports was down to 10 percent and Uruguay's was only 3 percent.[9]

The WTO's emergence ushered in a neoliberal food regime that embraced minimal inter-governmental standards. Under the WTO's Agreement on Agriculture in 1995, Philip McMichael notes that "states no longer have the right to food self-sufficiency as a national strategy."[10] WTO member states were required to accept food imports up to 5 percent of their total domestic consumption. CAP restraints on European agricultural overproduction were agreed to in 1992, as were reductions for US and European Community supports for export subsidies.[11] While global meat consumption has continued to increase, much of this increase has occurred in Asia. McMichael points out that "Asia, whose global consumer class outstrips that of North America and Europe combined, leads the livestock revolution."[12] In terms of feed grain production, the Americas, especially the United States, Brazil, and Argentina, now provide Asian countries, particularly China, with the corn and soybeans needed to support its livestock revolution. In 2007, China imported 45 percent of the world's soybeans used for animal feed. Europe has also become a net exporter of corn. While Asian corn production has been vastly improved due to adoption of hybrid seeds, particularly in China, it is nowhere near large enough to meet all its livestock feed requirements.[13]

## The Poultry Revolution

Consumer demand for meat around the globe after World War II increasingly focused on poultry. Indeed, its transformation into the meat of choice for Westerners and Easterners has been dramatic and was certainly unforeseen before the war. Throughout most of the meat-eating world until the post–World War II era, poultry was a rarely consumed meat, reserved for special occasions. Chickens themselves were quite common throughout the world, but eating more than a few pounds of chicken meat per year was unusual until after the mid-twentieth century. Over the course of the nineteenth century, American farmers increasingly raised chickens but did so primarily for the commercial value of their eggs. Eating chicken meat was a by-product of egg production, and generally only older chickens were consumed once they could no longer produce eggs. Among the world's various meat-eating cultures until modern times, perhaps the strongest identification with chicken meat was among Jews. In addition to their avoidance of pork as unclean, chickens occupied a special role among orthodox Jews on the day of *kapparah*, or atonement, on the second day before Yom Kippur. In urban Northeast America by the late nineteenth and early twentieth centuries, Jews were the main purchasers of chicken for Sabbath meals. Establishment of the chicken industry in the Delmarva Peninsula, which included the eastern shores of Delaware, Maryland, and Virginia, was largely due to this regional market demand. Chicken meat was even more expensive outside this region. For example, in 1914, at Chicago's exclusive Blackstone Hotel's restaurant, chicken cost more than crabmeat, prime rib, venison, and lobster.[14]

Before chicken production was revolutionized, increasingly industrialized egg production was underway in the United States at the turn of the twentieth century. Between 1899 and 1909, the value of poultry produced in the United States increased by almost 50 percent with most poultry production concentrated in the mid-Atlantic states, as well as Ohio, Iowa, Illinois, and Missouri. In 1910, 88 percent of all American farms kept an average of about 80 chickens per farm. Fifteen years later, the United States was the world's leading consumer of chicken meat at about 14 pounds per capita. Most consumption was still seasonal: chicken was primarily eaten during the fall months. Consumption rates in countries such as Canada, Denmark, and France were about half this amount. One way to consider the Republican Party's promise in 1928 to put a "chicken in every pot" was that it underlined just how rare chicken consumption was.[15]

William Boyd has emphasized that chicken's transformation into a commonplace meat combined "a complex mix of public research, private science, and business enterprise during the first half of the twentieth century."[16] This revolution created what many commentators refer to as the industrial (or technological) chicken. These developments began in the United States during the interwar period before migrating after World War II to Britain, Australia, and much of the rest of the globe.[17] Poultry science emerged during the 1910s and 1920s, focusing on issues related to "bacteriology, biochemistry, and the economics of egg and meat production."[18] The first step in making an industrial chicken was creating birds that could withstand intensive confinement. Adding vitamin D via cod-liver oil to their feed strengthened chickens' legs and allowed them to go without ultraviolet light in cramped indoor conditions. Creation of commercial hatcheries also helped to provide newly hatched chicks in sufficient numbers to growers. The federal government and private industry worked together in the 1930s to reduce death from a variety of diseases, such as pullorum, that afflicted confined birds. Nutritional studies helped to produce better feed from corn and soybeans, and studies of antibiotics demonstrated that using penicillin and chlortetracycline as feed additives sped up growth rates. Addition of vitamin B-12 also improved growth rates. Indeed, while antibiotics came to be widely used in all farm animals, the poultry industry relied especially on them. Breeding and genetic improvements also accelerated chicken's transition to industrial production. The meat-type chicken with broad and heavy breasts emerged due to the push for hybridization among chicken breeders. These efforts were helped by the fact that compared to other types of farm animals chickens' biological time lags in reproduction and growth rates were considerably shorter. The combined effects of these developments by the end of the twentieth century had brought the average time to produce a nearly five-pound broiler (the term used to describe young meat-producing hybrid chickens) down to fewer than two months and required less than two pounds of feed to produce each pound of broiler meat.[19]

During the middle third of the twentieth century, these scientific developments resulted in the transformation of the location and business structure of the US poultry business. In 1935, broilers were killed at sixteen weeks and required over four pounds of feed to produce each pound of meat. Not only were they expensive, but their production was confined primarily to the Delmarva Peninsula. However, during the Depression era, Jesse Jewell, John Tyson, and Charles Odell Lovette, all of whom would pioneer the vertical integration of the chicken industry after World War II, laid the groundwork for linking farmers through grower contracts, alliances with or development of feed companies, and experiments with new processing operations, as well as a location shift of broiler production to the Deep South, stretching from Georgia to Arkansas. During World War II, chicken was not rationed, and the federal government set its price well above cost of production, resulting in a tripling of broiler production between 1940 and 1945. Broiler meat finally surpassed chicken meat among American consumers in 1952.[20]

Over the next two decades, broiler production became the most thoroughly vertically organized segment of the meat industry with hatcheries, feed mills, and processing plants tied to chicken farmers producing birds under contract. By 1980, 99 percent of broilers and 90 percent of turkeys sold in the United States were grown directly by corporations or under production contracts with fewer than forty companies; and two companies—Tyson and ConAgra—dominated the production of chicken and turkey. Twenty-five years later, Tyson continued to dominate broiler production, with Pilgrim's Pride the nation's second-largest producer; but Jennie-O Turkey and Cargill Turkey Products had passed ConAgra as the top turkey producers. In 2014, Tyson and Pilgrim's Pride remained the two top broiler producers, and Butterball, Jennie-O Turkey, and Cargill were the three leading turkey producers in the United States. Eschewing whole chickens, which are lower in fat than beef, Americans more often consumed processed chicken pieces that they purchased in fast-food restaurants or in branded products at their retail supermarkets.[21]

The poultry revolution spread to Europe and other parts of the world after World War II. In some ways the rapid rise of chicken's popularity in Europe was even more astounding than in the United States. For instance, a 1954 survey in Britain found that one-third of households never bought chicken and that another one-quarter did so only at Christmas. At this time, London's Jewish community provided nearly all the demand for chicken. Geoffrey Sykes was the pioneer of cheaper chicken in Britain. Benefiting from chicken's exemption from rationing in the postwar period, Sykes and others developed productive breeds of chickens that were able to withstand the stress of concentrated living conditions. From the late 1950s to early 1960s, the average flock size in Britain grew rapidly from nearly all fewer than 1,000 in 1957 to over 20 percent between 1,000 and 5,000 by 1960. As in the United States, antibiotics were added to chicken feed to promote growth and to help the birds in confinement withstand disease, particularly coccidiosis. By 1960, the price of chicken in Britain was already lower than bacon, ham, beef, veal, mutton, and lamb. Unlike the United States, however, British poultry farmers did not pursue concentration at extreme levels. Bernard Matthews was Britain's leading innovator in turkey production. Over the span of the early 1950s through the late 1960s, he gradually built up production to over one million birds. His products attained strong market demand during the 1980s. Matthews purchased turkey businesses throughout Europe and New Zealand during the 1990s. In 1970, the British government outlawed wholesale use of antibiotics to promote growth, and much of the rest of Europe did the same during the 1980s. Retailers like J. Sainsbury were early promoters of healthier chicken products. As a result of its cheaper price, increasing convenience, and reputation for wholesomeness—especially in contrast to beef, which was plagued with the effects of the BSE crisis—chicken had become the most popular meat in the United Kingdom by the 1990s.[22]

Until about 1960, many Australians kept chickens in their backyard "chooks," and, like Americans early in the twentieth century, raised them primarily for eggs and not meat. According to Jane Dixon, for Australians "[r]oasting a chicken was an act of love, a symbol of a special occasion, a rare treat signaling a religious holiday."[23] The country's leading innovator in chicken production was Norm Thomas, a well-known bird breeder, who owned the Windsor Poultry Shop in Adelaide in the 1960s. From 1960 to the mid-1990s, annual average consumption of chicken increased from 11 to 61 pounds. Chicken and margarine were the fastest growing foods among Australian consumers from the mid-1970s to the mid-1990s. Although it took a bit longer for poultry to surpass beef in popularity compared to the United States or the United Kingdom, it did so early in the twenty-first century. As in the United States, the Australian poultry sector is dominated by very few firms, all of which rely on vertical integration and grower contracts.[24]

Starting in the 1970s Thailand was especially important among Asian nations in developing a poultry industry, particularly for marketing chicken to Japan. The commercial development of poultry through introduction of Western hybrid birds began in the mid-1970s. The industry adopted the full range of integrated commercial elements that allowed for poultry's development in the West, including links to feed-milling companies and contract farming. Charoen Pokphand (CP) Company introduced contract growing in the late 1970s, and by the late 1990s about ten to twelve companies controlled 80 percent of Thailand's broiler industry. In 2010, CP was the world's largest animal feed producer and one of the world's largest poultry exporters. Since then, CP has also expanded into pork production in several Asian countries and Russia. Chicken has become increasingly important in Thai diets, replacing much of the country's traditional fish, buffalo, beef, and pork consumption. Beginning in the early 2000s, at about the time that Thailand's industry was hurt by Avian flu, mainland China and Hong Kong became increasingly important poultry producers and exporters, though still somewhat behind both Brazil and the United States in the early 2010s.[25]

Although the poultry revolution's creation of rapidly falling prices contributed to chicken's growing popularity, other factors enhanced consumer acceptance around the globe. Chicken was increasingly marketed in ways that made it more convenient for consumers to eat—boned, cut into pieces, precooked, frozen, and packaged. Added to this was its status as a healthy—that is, less fat and cholesterol—type of meat. American health advocates Ancel and Margaret Keys's *Eat Well and Stay Well,* first published in 1959, noted explicitly that compared to beef and pork, "[c]hicken is very different. Chicken meat contains much less fat, especially if the birds are young."[26] In 1977, the *Dietary Goals for the United States* touted chicken as well: "Decrease consumption of animal fat, and choose meats, poultry and fish which will reduce saturated fat intake."[27] Although in the 1980s most Americans increasingly turned to products like McDonald's Chicken McNuggets, which were not less fatty than beef, the perception that chicken was a healthier meat choice has continued into the twenty-first century. In addition, most industrialized chicken has a neutral, even bland, taste that is readily flavored in ways that make it popular in a variety of foods. The popularity of many ethnic cuisines in which heavily flavored chicken serves a meat filler role makes neutral or bland taste acceptable. Many consumers judge chicken easier to eat than either beef or pork. Also, it is likely that many consumers have fewer ethical qualms about killing and consuming birds than mammals.[28]

## Changing Western Meat-Eating Preferences

The poultry revolution's effect on Western meat choices is clear and dramatic. In 1961, poultry was far behind beef and pork consumption in the major Western meat-eating countries. (See table 2 in appendix.) Fifty years later, it was the leading meat eaten in the United States, United Kingdom, Canada, Australia, Brazil, and nearly so in Uruguay and Japan. (See table 4 in appendix.) Although traditional meat-eating proclivities were still evident in 2011—for example, significant beef consumption in Argentina, New Zealand, and Australia and high pork consumption in Germany—other traditional meat-eating patterns disappeared in the post–World War II period. For instance, the UK's status as the world's premier beef-eating culture eroded after the war. Starting in 1961, the UK's dominant meat was pork. By the turn of the twenty-first century it was poultry. In the United States, preference for beef—itself a product of the early post–World War II period—remained in evidence through the last half of the twentieth century; but as in the case of the United Kingdom, poultry far outdistanced beef among American consumers' meat preferences by the early twenty-first century.

The other major meat-eating pattern evident in the West, especially after the war, was the transition in beef's status from a prestige meat to a convenience meat. The post–World War II increase in hamburger consumption was a product of the fresh beef revolution, but hamburgers were meant to be cheap and easy to consume. As convenience and fast foods garnered a progressively more important place in US consumers' food choices after World War II, hamburgers made up an increasingly significant proportion of Americans' beef choices. Between 1970 and 2005, ground beef's proportion of total beef consumed rose from about 25 percent to just over 50 percent. In 2015, the average American ate about thirty hamburgers a year, which was six more than they ate in 1976. At the same time, average amount of total beef consumed was almost half as much in 2015 as it was in 1976. Convenience, simplicity, and low cost made hamburgers the most important type of beef consumed in the United States after World War II. Moreover, these factors have also made hamburgers popular in Europe and much of the rest of the world. Today, Australians eat even more hamburgers on average than Americans, and people in the United Kingdom eat almost as many as Americans do, with the French, Russians, and Germans not far behind the British. Whereas meat, especially beef, was a signature part of meals among Westerners in the years before World War II, at the beginning of the twenty-first century, Westerners increasingly see meat as just one part of a much more varied diet in which convenience is perhaps the single most important factor in determining meat—as well as other food—choices.[29]

Brazil has been the most notable addition to the West's major meat-eating nations since World War II. The country's rise as a major producer and exporter of meat and feed grains has been rapid and dramatic during the period of the corporate food regime over the past generation. Brazil is also one of the few Western countries where beef consumption has increased substantially from 1961 to 2011, more than doubling in that half-century. (See tables 2 and 4 in appendix.) National policies stressed livestock development as part of the nation's overall economic development. In the late 1950s, Brazil had about 50 million cattle. Beginning in the mid-1970s, Brazil's cattle industry, particularly in the Amazon, started to grow dramatically. By 1975, Brazil's cattle numbered more than 90 million. Through the end of the twentieth century, much of this beef was marketed domestically. Then, between 2000 and 2010, Brazil's meat exports quadrupled. During that decade, Brazil's total cattle numbers rose from 170 million to 210 million. Its share of the world's meat exports rose from 6 to 16 percent, making Brazil the world's largest beef exporter and second-largest exporter of meat overall. Additionally, its increase in soybean production aimed primarily at meeting China's growing demand for feed grains and oilseed for its livestock industry has been equally dramatic. Brasil Foods, created in 2009 with the merger of Brazil's two largest poultry companies, Perdigão and Sadia, became the world's largest poultry producer, surpassing Tyson Foods. Exports made up almost half its total sales. JBS, a Brazilian meatpacking company known as Friboi until 2007, embarked on a rapid expansion phase and corporate takeover campaign beginning in that year that made it the world's largest meat producer by 2010.[30]

## Increasing Japan's Meat Consumption

Before the devastating effects of World War II, Japan's domestic meat supplies had grown somewhat fitfully during the 1920s and 1930s. Foreign supplies also added to Japan's overall meat supply. Particularly after Japan's invasion of Manchuria in 1931, meat was imported from China as well as Australia until the war began. After 1937, Japan's acquisition of Chinese territory allowed it to exploit not only meat supplies but vast amounts of grain. Typically, meat from China bound for Japan was slaughtered and processed in Qingdao, China, and then shipped to ports in Moji (near Shimonoseki) in Yamaguchi or Kobe for reshipment throughout the country. Because Japanese shipping became increasingly tied up with the fighting, however, meat shipments from China stopped in 1942. The war devastated Japan's developing animal livestock industry, as well as its people's access to meat. Not surprisingly, Japanese meat consumption at the end of the 1940s was less than it had been during the mid-1930s. On average, Japanese meat consumption in 1948–50 was just over 4 pounds per capita per year compared to almost 5 pounds per capita per year in 1934–38.[31] (See table 13 in appendix.)

Although meat became almost unavailable during the war—forcing consumers to turn to exotic alternatives including silkworm cocoons, worms, grasshoppers, mice, rats, moles, snails, snakes, or a powder made by drying the blood of cows, horses, and pigs—if anything, meat became even scarcer following the war. While the war and privation had contributed to this scarcity, many in the meat trade blamed government price controls. On May 22, 1947, the National Meat Industry Workers Association and Japan Livestock Association called for an abolition of the price control system at a rally numbering over two thousand held at the Shiba Public Hall in Shiba Park, Tokyo. Participants in the rally marched to the Diet Building and presented government officials with a petition signed by more than one hundred thousand people from across the country asking for the end of price controls. Despite these efforts, price controls continued. At the height of postwar inflation, one meat industry official estimated that the price of beef was more than double the rate of inflation for other products. The high price of meat sparked consumer protests that began with the Ōsaka Wives Association movement called "Cheap Meat for Working People" in July 1948. Pressure from other consumer groups in Tokyo and other large cities resulted in a new government act in August 1948 that superseded a similar act passed in 1946. These efforts seemed to have been ineffective, but clearly the fact that widespread protests about the price of meat occurred suggests that it had assumed a larger place in the Japanese consumer's diet.[32]

During the 1950s, Japan's meat associations embarked on several marketing and production paths to rebuild and enhance its consumer base. Two of the country's older cattle-rearing regions, Ōmi (today's Shiga Prefecture) and Ise (modern-day Mie Prefecture), received special emphasis in beef sales and promotions, particularly in Tokyo. Ise beef was marketed as *matsusaka* beef. Both Shiga and Mie Prefectures tried to capture the public's imagination for beef that was beautifully colored and marbled by claiming the animals had been massaged with rice straw on a daily basis. Another of the marketing promotions, beginning in 1950, was the creation of a meat festival to rival the traditional midsummer eel festival; eel was traditionally eaten during the hot, humid period of the year as a way to impart physical strength.[33]

Also in the 1950s, Japan embarked on a long period of economic growth and development now referred to as Japan's economic miracle. American meat products, like Hormel Company's Spam, a brand of preserved pork shoulder, gained a foothold. Some of the American occupation's economic programs may have contributed to this growth, such as the Farm Land Reform Law, passed in October 1946, which transferred a considerable amount of Japan's farmland from absentee landowners to former tenant farmers under the direction of local land commissions. The percentage of tenant farmers dropped from 46 percent in 1945 to 5 percent in 1950, when the land transfer was completed. The Japanese government continued its earlier policies of promoting livestock development through various incentives, such as subsidizing interest payments on loans for livestock purchases and importing cattle with loans from the World Bank. One government-sponsored livestock rebuilding project, the Livestock Holding Farm Establishment Program, subsidized introduction of beef and dairy cattle in areas of the country, particularly the mountainous areas of northern Honshū and Hokkaidō, which were not thought to be suitable for rice and other grain crop production. Although Japan's economic miracle may have had more to do with its shift to a high-tech export economy, its agricultural transition was critical in spurring important developments in terms of livestock production and meat consumption. Indeed, livestock production became an increasingly important sector of Japanese agriculture from the 1950s on. Numbers of cattle, hogs, and chickens all increased rapidly, and, through the 1960s, horse, goat, and even rabbit meat were important in increasing Japan's meat production, partly because they could be used in cheap processed meat products. One indication of the importance of these less prestigious meats was a scandal over the mislabeling of canned meat that supposedly contained beef. The *Shufuren* (housewives' federation) discovered that the meat was primarily whale or horse and persuaded the Japanese government to crack down on this practice of mislabeling.[34]

American livestock producers and their promoters in the US government saw tremendous opportunities for greater meat exports to Japan after the American occupation ended. American inspections of Japanese meat production facilities found numerous instances of waste and inefficiencies. While acknowledging the rapid growth in Japan's livestock supplies and meat production, they also highlighted that meat consumption remained low—just 7.7 pounds of meat per capita in 1958. American meat officials attributed this in large part to the high price of meat in Japan. In 1958, the average Japanese consumer allocated 43.8 percent of their average monthly household expenditures to food. But they directed just 11.2 percent of their food purchases toward meat. In the same year, the average American spent just 22.9 percent of their incomes on food. Of that much smaller relative expenditure, however, Americans on average spent 26.1 percent of their food dollars on meat. Nevertheless, during the late 1950s, the United States exported almost no meat, either fresh or frozen, to Japan. America was a significant exporter of tallow—used primarily in soap manufacturing—and calf and cattle hides to Japan, but otherwise played almost no role in Japan's meat market. Nearly all the imported meat, primarily beef since pork, mutton, and lamb imports were negligible, came from Australia and New Zealand.[35]

From the 1950s through the 1970s, Japan focused on controlling its own meat supplies through increased domestic production. (See tables 10–12 in appendix.) Food self-sufficiency was a major Japanese government priority, as was import protection for farmers. In particular, the Japanese government decided to focus on self-sufficiency in meat, dairy products, and rice, and allow imports to supply most of its animal feed needs. The government's focus on meat and dairy production was also part of its plan to boost farmers' income levels. Bolstering meat production and stabilizing livestock prices were aims of two important laws passed in 1961: the Agricultural Basic Law and the Law Concerning Stabilization of Livestock Prices. The former was aimed at stabilizing prices for agricultural products, rationalizing distribution, and reducing the gap between urban and rural incomes, particularly since urbanization was accelerating. The Agricultural Basic Law also encouraged farmers, primarily through import quotas and with grants and low-interest loans for new feedlots, to increase beef production through dairy beef. The surplus steers produced by accelerating dairy production could bolster cattle numbers traditionally supplied from *wagyū* steers.[36] The Law Concerning Stabilization of Livestock Prices supported the government's position that "livestock production was chosen as one of several key items for selective expansion and the stable promotion of the livestock industry soon became a major policy issue."[37] This law established the Livestock Industry Promotion Corporation, which not only focused on programs that subsidized, stabilized, and promoted domestic livestock but, starting in July 1966, governed the purchase and sale of imported beef.[38]

While not increasing its meat exports to Japan between the 1950s and 1970s, the United States played a crucial role in supporting Japan's increase in domestic meat supplies in the immediate postwar years, especially through its soybean exports. Japanese admission to GATT in 1955 helped it to develop this new trade relationship. Although consumed both by humans and animals, Japanese soybean imports increased fivefold between 1950 and 1959 with 94 percent of the imports coming from the United States in 1959. In that final year of the 1950s, one-third of American soybean exports went to Japan, making it the United States' largest recipient. The amount of crushed soybeans—used for animal feed—became more important in Japan than whole soybeans—used by humans—during the 1950s as well. By 1959, as its production of coarse grains, including rye, sorghum, barley, buckwheat, corn, millet, and oats were all in steep decline, Japan's import needs accounted for 5 percent of the world's total net grain consumption.[39]

With the Japanese economic miracle in full swing, Japanese consumers' desire for meat from the 1950s through the 1970s was met primarily by increased domestic meat production fueled by increased feed grain imports. (See tables 10–12 in appendix.) During the 1960s, meat consumption rose by just over 140 percent, while between 1963–64 and 1973–74, coarse grain imports tripled—from 4.6 million to 14 million metric tons. Japan's total average net grain imports of 19 million metric tons from 1973–74 to 1975–76 constituted two-thirds of the total amount of all grain imports in the developing world. About 12 million metric tons of the 19 million imported went to feed livestock. Despite the fact that Japanese consumption expenditures doubled from 1957 to 1963 and quadrupled between 1963 and 1973, the national savings rate increased by 5.5 times between 1963 and 1973.[40]

Because American grain sales to Japan declined in the early 1970s due to poor harvests and increased sales to the Soviet Union, Japan decided to reverse its declining coarse grain production for about a decade starting in the mid-1970s. Nevertheless, Japan's policy shift toward greater feed grain self-sufficiency was only a temporary change in the general post–World War II direction of the country's food policies. Japanese consumer demand for more and cheaper meat and other foods played an important role in encouraging the country's leaders to start to move away from its primary postwar emphasis on supporting domestic food production. Not coincidentally, during the early 1980s, average Japanese meat consumption surpassed the world average. (See table 13 in appendix.) Demand for beef produced primarily from domestic animals had pushed the retail prices in Tokyo up nearly sixfold between 1960 and 1977. Until the late 1980s, meat imports had increased but added only modestly to the total available supply between 1975 and 1985. Clearly,

Japan, like Western Europe, was a beneficiary of American (as well as Canadian, Australian, and Argentine) grain production and trade policies that focused primarily on affluent parts of the world that wanted to eat more meat. During the same years that Japan made some effort to increase its own livestock feed grain production, the country was debating a fundamental shift in its post–World War II meat production strategies. Partly in response to an increasingly complicated and multifaceted trade relationship with the United States in particular, Japan gradually moved its position from attempts to maximize its agricultural self-sufficiency to a position of greater willingness to import more food, including meat from the United States and other countries. In fact, imports of poultry products had already been liberalized during the 1960s, and pork products had been liberalized during the 1970s. These policies became increasingly evident in the beef trade negotiations between Japan, the United States, and Australia between 1978 and 1991 that resulted in substantial increases in Japan's beef imports from both countries.[41]

Japan's decision to increase its beef imports during the opening years of the Heisei era (1989–present) is just one of many elements that have influenced Japan's contemporary meat culture. While liberalizing beef imports opened Japan to more trade with Australia and the United States, it also resulted in more joint ventures and Japanese direct investment in livestock and meat production in Australia; Nippon Meat Packers and Itoham grew especially due to these developments. By 1993, just two years after Japan's beef import agreement with the United States had been fully implemented, Japanese corporations owned twenty-one meatpacking plants in nine US states. During the 1990s, Nippon Meat Packers became the world's fourth-largest meatpacking firm, particularly because of its establishment of plants in several countries. In 1990, 15 percent of total beef production in Australia was controlled by some degree of Japanese ownership, particularly Nippon Meat Packers and Itoham. As Japan imported more meat and food generally, it also increasingly adopted Western styles of food consumption, including fast food. Nevertheless, probably as much meat continued to be consumed through older Japanese foods, such as curry rice, *tonkatsu*, and *sukiyaki*—which in the early twentieth century had themselves been considered Western-style foods—as through more modern Western-style foods.[42]

Some indication of the types of meat that Japanese preferred is evident in popular cooking *manga* (comics), such as Kaneko Nobuo's

*Tanoshii Yūshoku* (*Fun Dinner*), published in 1990. Kaneko acknowledged that hamburgers and croquettes were already popular, but he also advocated a wide array of other meat dishes. *Fun Dinner* is divided into sections that focus on four types of meat dishes: beef, pork, chicken, and ground meat. In terms of beef, several recipes incorporate Japanese cooking styles, flavors, and emphases using thin slices or small pieces of meat, including wet preparations such as stews and soups as well as hamburger and bacon *maki* (sushi) and cold beef with *wasabi* (green horseradish). The pork, chicken, and ground meat sections are similar, with recipes that incorporate the meats into soups and steamed curries and recipes for meats combined with various vegetables. The ground meat section includes several recipes using sausages, an item that seems to have caught on early in Japan's transition to meat and stayed popular ever since. Kaneko also includes recipes that focus readers' attention on particular regions of Japan and heighten interest for nostalgic purposes. For instance, he highlights *jibu-ni*, a simmered chicken dish famous in Ishikawa Prefecture; *chikuzen-ni*, another simmered chicken dish famous in northern Kyushu; and Okinawa soup with pork fillets. Harking back to a much earlier era of meat promotions, Kaneko also felt the need to highlight how meat makes people healthy, explaining his views in a way that eerily echoes Meiji era promotions. Similarly, he maintains that those who do not eat meat lack stamina and are less able to withstand the heat and humidity of Japan's summers. Meat, according to Kaneko, is also a medicine, which is, of course, another traditional, but apparently, still compelling argument for meat eating in Japan.[43]

As was the case in the United States and elsewhere in the West, Japan's meat consumption accelerated due to the rapid acceptance, especially among young people, of fast-food hamburgers and other cheap meats, particularly since the 1970s. While the first burger shops opened in Nagasaki and Tokyo in 1951, Japan did not have a burger chain until Dom Dom Burger began in 1970, followed by McDonald's in 1971. Most Japanese had looked unfavorably upon fast food until the opening of chain restaurants. Part of the negative associations with fast-food burgers was standing and eating (*tachigui*) and eating with fingers (*tezukami*). McDonald's in Japan, founded by Fujita Den, helped to break down these negative associations among young people. In fact, by 1984, McDonald's ranked first in Japan in total restaurant sales. Japan-style McDonald's food items, such as the Teriyaki MacBurger, also helped to popularize fast food. Mos Burger, founded in 1972, was even more innovative in developing Japan-style fast-food items, particularly after 1987 when it introduced the Mos Rice Burger, with a bun made from rice instead of wheat flour.[44]

By the beginning of the twenty-first century, Japan had become the largest US export market for beef. These exports totaled $1.3 billion and accounted for 37 percent of American beef exports. US beef supplied much of Japan's grain-fed beef, which was preferred for the restaurant market for items like chuck roll and short ribs and had been used in *gyūdon* (simmered beef served on top of rice) and *yakiniku* (grilled meat). *Yakiniku* can also include pork and chicken. The beef in *yakiniku* is prepared quite similarly to that in *yakinabe*. Specifically, Japan had imported two main types of beef from the United States: boneless chilled or frozen cuts and edible offal, particularly beef tongue used extensively in *yakiniku*. In December 2003, Americans inspectors reported that bovine spongiform encephalopathy (BSE) had been discovered in a cow in the United States. Although BSE had actually been discovered in a Japanese cow in August 2001, the American report resulted in Japan's decision to close its market to American beef exports. Japan then largely turned to Australia and New Zealand to supply more of its beef. These countries had already dominated Japan's grass-fed beef market. Between 2004 and mid-2006, when Japan nearly completely banned US beef exports, American authorities estimated that their export losses totaled $5.7 billion. In 2006, the American and Japanese governments agreed to a Beef Export Verification program to certify BSE-free American beef exports to Japan. A key provision was that exported beef come from animals aged twenty months or less. Japan was given authority to audit US beef plants as well. Beginning in July 2006, American beef exports resumed and recovered somewhat, though by 2007, American beef exports to Japan totaled just $244 million and were a distant second to Australia's $1.8 billion and just ahead of New Zealand's $177 million.[45]

The BSE threat raised the already high level of Japanese concern about food security to even greater heights. As part of its screening for BSE, the Japanese government decided to test all domestic cattle at the time of slaughter. The program cost an estimated $10 billion in 2008. While even many Japanese authorities believed that universal testing was not necessary, this testing regimen helped to restore consumer confidence in the domestic beef supply. Doing so perhaps inadvertently helped to propagate a negative view of foreign, especially American, beef products. A 2006 Japanese survey found that 80 percent of the country's consumers worried about the safety of imported beef. A 2007 survey indicated that Japanese consumers preferred Mos Burger over McDonald's.[46]

While Japanese opinion of US beef safety improved considerably after 2008 and polls even found that McDonald's surpassed Mos Burger in popularity among burger chains in 2009, Japan's imports of US beef still remained limited in 2010 compared to 2003 levels. US Secretary of Agriculture Tom Vilsack visited Japan in April 2010 to press for further steps to reopen the country to American beef. Even though many more cases of BSE had been discovered among Japanese cattle than American (thirty-six compared to three), Japanese consumers remained suspicious because US authorities did not test all cattle for BSE. Shortly after Vilsack's visit, foot-and-mouth disease was discovered in cattle in Miyazaki, resulting in the destruction of nearly three hundred thousand animals between April and July 2010, when the state of emergency was lifted. The disease hit the *wagyū* cattle that were prized for export markets. Imports of US beef nearly doubled between 2009 and 2010.[47]

## Increasing China's Meat Consumption

Although Japan's meat consumption increased steadily during the 1960s and 1970s and reached world average levels during the early 1980s, China's meat consumption was still miniscule at the beginning of the 1980s. Edward Rada noted in 1983 that "[t]he shortage of meat makes China virtually a vegetarian country with the essential amino acids coming from a variety of vegetables, legumes, and grains."[48] Indeed, between the 1930s and 1976, Chinese per capita food and grain production had fallen between 12 and 17 percent. Using data from John Lossing Buck's studies of rural Chinese diets from the late 1920s and early 1930s as a baseline, animal food availability by the late 1970s was only half of what it had been in 1929–33. During this half-century period, China not only had suffered through the dislocations caused by fighting with Japan during World War II that overlapped with a long period of civil war, but also had experienced the world's largest and most severe famine in world history during the years from 1958 to 1962. Both animal and plant food supplies started to increase during the late 1970s, but most protein sources were produced for on-farm consumption, which, of course, made sense in a nation that was still largely rural in the early 1980s. Agriculture's share of China's employment in 1980–83 was 75 percent.[49]

While informed observers such as Vaclav Smil noted the improvement in food availability in China over the late 1970s and early 1980s, Smil was still doubtful in 1986 that China's grain-based diet would change much over the next generation. "Realistically, the best we can expect is a doubling of the Chinese animal food intake over the next generation. Animal food would account for at least 12, possibly up to 15 percent of all food energy, still below the current Japanese level."[50]

Smil's predictions fell far short of reality in terms of predicting China's extraordinary dietary revolution. Reform of the country's agricultural sector became one of the linchpins of Chinese economic transformation. After Mao's death in 1976, China began the process of socialist modernization that would place it on the road to capitalism beginning during the late 1970s. Beginning in 1977, local party officials encouraged peasants to pursue all available opportunities to increase their incomes, including animal husbandry. Officials believed that peasants who ate well would be less likely to undermine government directives. Although decollectivization and household agriculture became commonplace in the late 1970s, the government started to dismantle the commune-based collective agriculture starting in 1980 and then abolished it and removed government administration from agricultural management between 1983 and 1985. Confined animal feeding operations (CAFOs) were introduced in 1979. Domestic meat price controls were liberalized in 1985. Between 1979 and 1984, agricultural production grew by 6.7 percent per year. Kym Anderson and Rodney Tyers's analysis of China's economic growth in 1985 pointed out that per capita meat consumption had increased by two-thirds from the 1970s, and they predicted that meat production would continue to grow between 4 and 5 percent per year through 1995. According to Lester Brown, Chinese meat consumption actually increased by 400 percent between 1977 and 1994.[51]

At the start of the 1990s, China's growing population of urban residents and its more affluent rural population had increased their consumption of meat substantially and also prompted a distinctive shift toward production of more animal fodder and less rice and wheat for human consumption. In 1992, the average resident of Shanghai ate over 132 pounds of meat, fish, and eggs per year, which was twice the national average and three times the average of poor residents in the country's interior. The correlation between urban residents and higher meat consumption, though arguably more dramatic in modern China, mirrored the same relationship that had existed in Western nations since the nineteenth century. Like the Japanese and South Koreans before them, the more affluent Chinese consumers—with urban residents' incomes doubling and rural residents' incomes tripling in the 1980s—wanted a richer and more diversified diet with more protein and fat content. During this period, China relied on domestic production for its animal fodder rather than turning to imports as Japan had done. Total percentage of agricultural land devoted to grain production fell slightly from 80.1 to 76.5 between 1980 and 1990. The average annual percentage increase in meat production between 1986 and 1990 was 7.7, significantly higher than Anderson and Tyers's predictions made in 1985 for this period. Indeed, compared to other East Asians, Chinese consumers in the 1980s devoted higher proportions of their incomes to meat than even more affluent East Asian neighbors. In 1990, average Chinese spent 27 percent of their income on meat.[52]

During the 1990s, the Chinese government encouraged CAFO production that resulted in continuing high growth rates to support increased meat consumption. During that decade, China surpassed the United States as the world's leading meat producer. This government emphasis required massive new transportation projects that helped to better integrate the inland and coastal regions. Nevertheless, at the start of the twenty-first century, China retained a regional self-sufficiency focus in agriculture that has been difficult to overcome despite national government policies and projects. As chapter 8 explains in more detail, this rapid intensification came at significant environmental costs. Part of the reason for the environmental problems, however, is the fact that China's agricultural sector retains a significant small-scale character. During the 1990s and into the 2000s, China's livestock and meat industries especially retained a household economic emphasis that is nearly forgotten in the West. In 2003, 102 million households raised from one to nine pigs, and only 4.5 percent of farms were considered intensive. Nevertheless, CAFOs established an important foothold in China's livestock industry during the 1990s. They accounted for 15 percent of pork production and 40 percent of broiler production by the end of that decade. Between 1996 and 2011, the number of broiler farms decreased by 75 percent. At the beginning of the twenty-first century, just over 6 percent of livestock growers produced nearly 56 percent of the country's animal products.[53]

China's national government's efforts encouraged freer, market-oriented livestock and beef industries while it also emphasized programs that managed this development. A central concern was diversifying meat product availability. In the early 1980s, pork constituted about 82 percent of the available meat in China. Chicken provided about 7 percent, beef 2 percent, and sheep 2 percent of consumers' meat. Although this new emphasis started earlier, the Chinese Ministry of Agriculture issued a policy document in 1999 that approved livestock sector restructuring to give dairy, fine wool, beef, sheep, and poultry priority. The beef sector received considerable attention due to consumers' desire for more fresh beef and a higher quality product. As it symbolized during the nineteenth century in the West, beef eating was an indicator of prosperity for late-twentieth-century Chinese consumers. Urban consumers also generally saw beef as a healthy alternative to pork, since, in terms of how it was sold, it was lower in fat. Since the 1980s, the national government encouraged more specialized and commercialized beef cattle development. Beginning with the Straw for Beef (later renamed Straw for Ruminants) program, the government hoped that encouraging more beef consumption would preserve more domestically produced grain as well as provide environmental benefits by using straw that was traditionally either burned or used in the water-polluting paper industry. Both cattle-raising and the beef industry gravitated from the country's western plains, where it had long been situated, toward the central plains, where a much higher proportion of the population was found. Even with transportation improvements, cattle growers had little incentive to engage in long-distance marketing, especially given that information on cattle prices and cattle classification and grading systems was underdeveloped. Consumer concerns about a more hygienic and safer product were partially addressed through the implementation of the Animal Epidemic Prevention Law that took effect in January 1998. Cattle slaughter was restricted to specific places where it could be inspected. Before then, the vast majority of cattle were slaughtered by households, to be sold in nearby markets. Between 2002 and 2010, per capita beef consumption rose 31 percent, the fastest rate of increase in the world. Sheep meat received less attention than beef, but the location of its production also shifted from the western plains to the central plains. Much of the product went to Muslim and Mongolian populations, but mutton and lamb have gained some popularity among other Chinese in recent years. By 2004, pork consumption fell to about 65 percent of Chinese meat consumption, while chicken increased to 13 percent, beef was up to 9 percent, and sheep was at 3 percent.[54]

Passage of the China Animal Husbandry Law, revised in December 2005, continued the focus on modernizing the livestock and meat production sector and providing higher quality meat products. This law preceded a series of disease issues in the pork and poultry industry during the mid-2000s that further encouraged the push toward intensive livestock and larger-scale meat production. In the 1980s and 1990s, government-owned slaughterhouses proliferated, many under the name of General Food Company, but primarily produced low-quality products that were often exported to Russia. Since then modern slaughterhouses replaced many of the General Food Company plants and increasingly produced higher-quality meat for more affluent Chinese. Most of these slaughterhouses are quite small by Western standards. Although the urban premium beef market has grown considerably since the 1990s, most Chinese at the turn of the twenty-first century still preferred to purchase their beef and beef edible by-products—especially tripe—in the traditional *nongmao* (rural trade or agricultural produce) markets for "wet" preparations, including stewing, stir-frying, and hot pot dishes. In addition to the ongoing concerns about environmental effects of China's rapidly growing livestock and meat production sector, many in China and elsewhere are alarmed by the health impacts of Chinese consumers' rapid transition to a more Westernized diet. This issue is examined in greater detail in chapter 9. Particularly since the 1990s, China's continued growth in livestock and meat production—now constituting one-third of total world production—has been propelled by imports of both meat and animal fodder and feed, much as Japan's supplies were a generation earlier. Some Chinese agricultural experts hope this trend can be reversed through adoption of more ecologically friendly agriculture based on traditional Chinese husbandry methods. Emblematic of China's new status as the world's fastest growing meat-eating country was the acquisition of the largest US pork producer, Smithfield Foods, by China's largest meat company, Shuanghui International Holdings Limited, in 2013 for $4.7 billion. Renamed WH Group in 2014, it is now the world's largest pork producer.[55]

The intersection of new trade patterns, state support for meat production, and shifting meat preferences are all prominent in the evolution of the political economy of meat since World War II. The astounding rise of Japan and China to near Western levels of meat consumption demonstrates how the groundwork for consumer acceptance of meat consumption before the war was increasingly met by rising feed grain imports from the United States through the last half of the twentieth century and then, increasingly, from Brazil during the early twenty-first century combined with state promotion of meat production and adoption of the scientific and entrepreneurial elements that have transformed poultry as well as pork into industrialized and commonplace meat commodities. Crucially, this changing political economy of meat has also resulted in a widening array of consumer concerns and resistance about its effects on workers, the environment, and human health. The leveling off of meat consumption in the West during the end of the twentieth and beginning of the twenty-first centuries suggests that many consumers are ambivalent and even wary of meat in a way that was not true during the heyday of rapidly rising meat consumption from the end of the nineteenth through the mid-twentieth century. Around the world, many segments of societies, including national governments, are increasingly conveying messages of moderation and even concern to consumers. The following three chapters of part III address consumers' increasing meat resistance.

# 7

## Meat Workers as Outcastes

FOOD SYSTEM WORKERS AROUND THE WORLD FACE widespread economic, ethnic, and gender injustices and inequalities. In the United States today, in addition to being poorly paid, they are also disproportionately nonwhite, with the nonwhite workers earning much less than even their white counterparts. Food system workers often lack labor union representation and, as a consequence, face dangerous workplace conditions. Meat workers share these problems with other food system workers, but they also face unique workplace hardships and cultural discrimination associated with their role in an industry that kills and disassembles animals en masse. This cultural stigma is longstanding in many parts of the world. Indeed, as meat has become an increasingly significant global dietary component, negative public perceptions of meat workers are likely worse than ever. This chapter will chart the most salient longstanding prejudices against meat workers, highlight countervailing trends that developed as states supported meat consumption, and then explain how antipathy toward meat workers reemerged during the late twentieth century as state protections for meat production eroded and a corporate food regime dominated by TNCs assumed greater prominence. As part of this latest development, the rise of a modern animal rights movement has focused more attention on the ethics of animal handling and slaughter, with which immigrant populations are now associated. This element may also play an increasing role in limiting people's meat consumption.[1]

Cultural and legal traditions that ostracized meat workers are especially longstanding in South and East Asia. Although meat workers in the West were not historically classified as outcastes and subjected to social and legal prejudices as they were in Asian countries like India, Korea, and Japan, important European traditions of ostracizing butchers proscribed them from specific civic duties, such as jury trials, and deep prejudices about butchers as inherently cruel were prevalent. This is a dominant theme, for instance, in much of the Romantic Era literature that either focused or touched on humane treatment of animals. Debates over *shechita*, or kosher butchering, starting in the late nineteenth century, not only affected treatment of Jewish butchers but were also part of growing disputes about the status of all Jews in Germany.

Nevertheless, as states' roles expanded in protecting consumers from dangerous meat products during the era of the first and into the second food regime, particularly through public ownership of slaughterhouses and support for unionization, perceptions of meat workers improved, and meat workers also made significant economic gains. However, starting in the second food regime and accelerating during the transition to a corporate food regime, countervailing trends, such as the movement of animal slaughter locations from urban to rural areas and increasing reliance on nonwhite and immigrant labor, have helped to reinvigorate old prejudices: meat workers—who now also face heightened scrutiny by the rising numbers of people concerned about animal rights and protections—are even more deeply marginalized and demeaned. Reliance on foreign-born meat workers became increasingly significant in the United States, several European countries, and Australia during the latter decades of the twentieth century and into the twenty-first century, as TNCs assumed dominance in the free trade and deregulation movement ensconced in the WTO. This chapter analyzes the apparent paradox that, as meat consumption has increased, more people around the world express greater hostility toward those who make their meat.

## Traditional Prejudices against Meat Workers

The essential paradox at the heart of how meat workers have become increasingly discriminated against across much of the globe in the modern era echoes that which for so long outraged the "untouchable" status groups in South and East Asia. That is, how can the broader population discriminate against a particular group for doing something that the rest of society benefits from? Herbert Passin was one of the early American commentators on the notion of untouchability in Asia. Comparing the low castes of India, the *paekchŏng* of Korea, the *vagyappa* of Tibet, and the *eta-hinin* of Japan, Passin identified three elements in common among these groups: widespread perception that such groups were hereditary castes; shared religious perceptions that pariah status reflected physical and ritual pollution and defilement; and pervasive beliefs that their extracaste status equated to an absence of virtue. In the case of Japan, Tokugawa-era legal codes defined the *eta*, as part of the outcaste *Buraku*, as a hereditary status group. Indeed, many Japanese long into the twentieth century believed that the *eta* were a separate race. The two major religious belief systems, Shinto and Buddhism, castigated the *eta* as defiled: Shinto emphasized the importance of ritual cleanliness, which the *eta*'s handling of animal products violated, while Buddhism discouraged meat eating and killing of animals and reinforced Shinto's notions that blood and death were ritually polluting. Together, Shinto and Buddhism stressed that the *eta* were immersed in blood, death, and filth. Because of these religiously reinforced beliefs, Japanese widely considered *eta* without moral virtue. As part of a broader category of *senmin*, or humble people, that included criminals, these beliefs were powerful and pervasive. One of the many indications of this during the Tokugawa era was classifying the *eta* as *hiki* (animal). In the mid-twentieth century, these characterizations were evident, for instance, in Sumii Sué's reference to *eta* being shunned as "Four-Fingers" in *The River with No Bridge*. Long after the early Meiji era edicts officially abolished the status system, *Buraku* faced employment, housing, and even marriage discrimination. After World War II, *Buraku* still often lived in the poorest areas of cities or villages, often lacked access to water and indoor plumbing, and received fewer educational opportunities. Discriminatory family registers classified them as *Tokushu Buraku* and resulted in social prejudice in terms of their marriage possibilities. As late as the 1990s, *Buraku* still reported that other Japanese often scorned them as *yottsu* (four legs) or *dōbutsu* (animal) and insulted them with the hand signal of four fingers pointed downward.[2]

Prejudices of various types existed toward meat workers in the West before industrialization. Before the nineteenth century throughout Western Europe and the Americas, most slaughtering and butchering of animals for human consumption took place in urban areas. Tanneries, which created noxious water run-off and horrific odors, were often classified as a nuisance trade but were also often located in or quite close to urban centers. In addition to their association with nuisance trades, the general assumption was that butchers were violent and brutal in character. For instance, John Locke's acceptance of this perception was implicit in his admonition for teaching children to be concerned about the welfare of all living creatures. He noted that butchers had long been excluded from "juries of life and death" because "they who delight in the suffering and destruction of inferior creatures will not be apt to be very compassionate or benign to those of their own kind."[3] The assumption that killing and cruelty toward animals would be reflected in human behaviors toward other people is evident in English artist and social critic William Hogarth's famous series of four engravings, *Stages of Cruelty*, from 1751.[4]

Traditional sentiments in the West about meat workers' dubious ethics and their contributions to urban pollution combined with expressions of contempt for their treatment of animals during the nineteenth century. The English public grew especially concerned about the issue of butchers and animal cruelty. By the end of the nineteenth century, greater concern for animals' quality of life and death, which accompanied the shift in slaughtering operations outside concentrated urban areas, triggered the animal protection and humane slaughter movements. Increasingly, the underlying sentiment seemed to be that animal death for the production of meat was necessary and perhaps inevitable but should nonetheless be merciful.[5] For instance, in 1816, James Plumptre, vicar of Great Gransden in Huntingdonshire, published *The Experienced Butcher*, which highlighted the widespread and hypocritical abuse heaped on the butcher's profession. Plumptre advised butchers "that, in slaughtering animals, every thing [*sic*] should be so ordered, as to give the animals the least pain possible . . . and the greatest cleanliness should be observed both in respect to the appearance and the smell." At the same time, he chided butchers to let their "other occupations and amusements be of a nature to counteract any disposition to cruelty," particularly avoiding "bull-baitings, cockfightings, racing, shooting, and hunting."[6] Samuel Coleridge's protégé, the English essayist Charles Lamb, wrote in his essay "Dissertation upon Roast Pig," published as part of *The Essays of Elia* between 1820 and 1825, that meat eating was a guilty pleasure that should not be confused with human sustenance.[7] Henry Bergh, founder of the American Society for the Prevention of Cruelty to Animals, was a key advocate of "necessary killing." He believed that people needed to eat meat but that the animals' deaths for this purpose should be quick and merciful.[8]

Emphasis on necessary but merciful animal killing resulted in increasing focus on the tools and techniques of slaughtering operations. As Ian MacLachlan has explained, three traditional techniques were used in the nineteenth century with cattle: stunning with poleaxes; felling with a nape stab; and *shechita*, the process of bleeding out—exsanguination—used particularly in Jewish ritual slaughter. Reformers saw problems with each of the techniques in terms of potential animal pain and suffering. An expert might wield a poleaxe and stun an animal with one blow. But often it took several blows before a steer was knocked senseless. The alternative to use of the poleaxe was the use of a short, double-edged dagger that would be thrust into the base of the steer's skull thereby severing the spinal cord. This process would normally cause unconsciousness and paralysis. Referred to as a "nape stab," reformers argued that often the unconsciousness and paralysis was not immediate. The third technique, *shechita*, was arguably the most contentious of the three. It required cutting the throat nearly down to the animal's spine. In orthodox Jewish ritual slaughter, the operation was performed by a *shochet*, who was trained to use a *chalef*, a razor-sharp straight blade about eighteen inches long with a squared-off end. Much of the concern with this technique was in terms of the way in which the animal was thrown to the ground to position it properly for throat cutting. If done correctly, a steer remained conscious for about twenty seconds but arguably without feeling pain.[9]

Among humane slaughter advocates, Jewish *shechita* was generally considered the most brutal of the various slaughter techniques. While the percentage of orthodox religious practitioners dropped rapidly in Europe during the mid- to late-nineteenth century to only about 15 percent of all Jews, organizations like the Royal Society for the Prevention of Cruelty to Animals and the Humanitarian League leveled increasing criticism on the practice of *shechita* and its alleged cruelty. Because of the orthodox prohibition against consuming any blood, *shechita* was meant to result in complete bleeding out by severing the large blood vessels in the neck. Despite arguments by physiologists and other scientists that the rapid removal of blood resulted in meat that tasted better, was more nutritious, and was more easily digestible, the allegations that ritual slaughter was a form of animal torture were difficult to eradicate from discussions. Many of the complaints were also tied to anti-Semitic literature and promotions.[10]

Arguably the most significant work published in the nineteenth century on the issue of animal treatment was the English social reformer Henry S. Salt's *Animals' Rights*, which first appeared in 1892. Salt's book both captured the concerns of that century's animal protection and humane slaughter advocates and also foreshadowed many of the arguments made in the modern era regarding animal rights, animal liberation, and the intrinsic value of animal life in general. Indeed, the use of the term "rights" in Salt's book signaled a new perspective about animals' lives that would be examined empirically and defended increasingly in the late twentieth century in works dealing with animal consciousness, such as Peter Singer's *Animal Liberation* and Tom Regan's *The Case for Animal Rights*. Salt elevated the level of thinking and concern about animals' lives above that of mercy and sympathy to a position of inherent worth more comparable to that traditionally accorded only to human lives. Salt's perspectives encompassed not only issues dealing with basic principles but treatment of both domestic and wild animals, animal "sports," fur and feather industries, animal vivisection, and also the slaughter of animals for food. Regarding animal slaughter, Salt made two main points. The first was that the underlying system involved in animal slaughter had become more, not less, cruel over time due both to the methods used to transport animals to slaughterhouses and the methods practiced in, especially, private slaughterhouses. Indeed, Salt suggested that some cruelties due to the latter issue would be improved if public slaughterhouses replaced them.[11]

One of the sources that informed his perspectives on the increasing cruelty of animal transport was Samuel Plimsoll's *Cattle Ships*, published in 1890. Written to support Plimsoll's efforts for legislation that would curtail the live cattle trade, the book is filled with descriptions of both animal and human suffering aboard the cattle ships used to transport livestock from the United States to Britain during the late nineteenth century. Regarding the animal suffering, for instance, Plimsoll wrote, "The arrangements for stowing a cargo of cattle are usually of the cheapest, flimsiest, and most temporary description. The suffering inflicted on the wretched brutes is horrible. In winter they freeze on the upper deck; in summer they swelter in the hold, and perish for want of air." Probably equally influential in Salt's notions of the cruelty of the people involved in the animal transport business were Plimsoll's descriptions of intentional abuse. Although Plimsoll also devoted considerable attention to the ways in which the cattle shipping trade harmed human lives, his description of animal torture are horrific. Because the men who worked on the cattle ships were paid on the percentage of cattle that survived the trans-Atlantic journey, they often did all that they could to prod the animals into leaving the ships. Plimsoll quotes one observer: "On several occasions I saw the men pour paraffin oil into [the cattle's] ears, which, as soon as it reached the brain, caused the poor brutes to fairly shriek with pain. Occasionally the ears were stuffed with hay, which was then fired; while in many instances the tails were snapped in the endeavours of the cattle-men to force the animals that had lain down from sheer exhaustion to regain their feet."[12]

Salt's second point encompassed the heart of the underlying paradox about making meat workers into outcastes.

> [T]he feeling of repugnance caused among all people of sensibility and refinement by the sight, or mention, or even thought, of the business of the butcher are [*sic*] also largely on the increase; so that the details of the revolting process are, as far as possible, kept carefully out of sight and out of mind, being delegated to a pariah class who do the work which most educated persons would shrink from doing themselves. In these two facts we have clear evidence, first that there is good reason why the public conscience, or at any rate the humanitarian conscience, should be uneasy concerning the slaughter of "live-stock," and secondly that this uneasiness is already to a large extent developed and manifested.[13]

In a single paragraph, Salt encapsulated the results of the separation of slaughtering from the provisioning of meat in terms of "humanitarian conscience." As Salt puts it, the public had relegated this duty to a "pariah class" that was kept increasingly out of sight and mind. Of course, Salt's larger point in the book was that this humanitarian conscience should evolve through education and legislation—his two major suggested remedies—into putting an end to all forms of animal cruelty. But in terms of the evolution in thinking about meat workers, the description of meat workers as a "pariah class" from whom the public "shrinks" is quite telling about how meat workers would be increasingly castigated and scorned around the globe.[14]

One of the sources that informed Salt's perspectives on the sensibilities of the so-called pariah class was H. F. Lester's pamphlet for the Humanitarian League titled *Behind the Scenes in Slaughter-houses*, published in 1892. In it, Lester made the case that most people "treat the operative slaughterer . . . as an unclean creature, a pariah of society." He contended this is so because "persons who have attained to any degree of education or refinement" are naturally revolted by the "act of killing a fellow-creature." Lester went on to note that "the ranks of slaughtermen are habitually made up from the dregs of the population, persons in whom one could hardly expect to find the sentiment of pity strongly developed." Yet he noted that even among the ranks of butchers were many who were "much too good for the business" but who felt they were compelled to do it because of the good wages and the lack other types of employment or who actually disliked the work but conceded that "'people will have meat.'" Lester also argued that meat eaters cannot simply pass off their participation in it without consequence. "But, happily there are among those who eat flesh many whose consciences are touched with the thought of the consequences which the practice entails both on men and animals." For both the less degraded butchers and the meat-eaters of good conscience, Lester made the case for humane methods of slaughter.[15]

## State Support for Meat Workers

Although prejudices against butchers and meat workers continued into the twentieth century, particularly against ethnic newcomers to the industry, state involvement in promoting and improving the quality of meat products helped to ameliorate some of these animosities. Although not as significant a goal of state support for meat consumption as increasing meat safety and improving meat trade, governments in the West and Japan nevertheless advocated improvements for meat workers' economic and social status in a variety of ways from the late nineteenth century through the 1970s. In Japan, the state increasingly intervened after World War II to rectify long-held social prejudices against the *Buraku*, though to be sure much of this intervention was a reaction to political mobilization of pressure groups, especially *Suiheisha*, a political movement that culminated in the formation of the Buraku Liberation League in 1955. In the West, states supported labor union and other economic protections that benefited meat workers.[16]

After the Meiji government issued the Emancipation Edict in 1871, little changed in terms of discrimination against the *Buraku*. Beginning in the 1890s, they pursued political activity; and, in March 1922, their representatives held the inaugural meeting of the *Zenkoku Suiheisha*, or National Levelers Association, in Kyoto. Until suppressed in the 1930s, *Suiheisha* organized political activities and ran candidates for government who raised *Buraku* concerns. Following formation of the *Buraku Kaihō Dōmei*, or Buraku Liberation League (BLL), in 1955, the next twenty years saw important state interventions on behalf of the *Buraku* and in terms of labor relations in Japan's meat industry. Intervention was particularly notable in the public sector meat production centers because *Buraku* were disproportionately employed in them.[17]

Largely in response to the BLL's formation and organizational efforts, government programs and the labor union movement attempted to tackle *Buraku* discrimination. The post-occupation government in 1953 began to make money available to redress *Buraku* discrimination. After the BLL and various left-wing organizations continued to pressure the national government for further reform, the Law on Special Measures for *Dōwa* (Outcaste) Projects was passed in July 1969. From 1969 to 1994, among the various levels of government in Japan an estimated $134 billion was spent on projects targeting the improvement of *Buraku* housing, health, and education. Other programs, particularly in Ōsaka Prefecture with its historically large *Buraku* population, continued into the twenty-first century.[18]

The BLL was also instrumental in the 1960s and 1970s in supporting efforts to reform abuses of packinghouse workers, especially in the large public sector packing plants in Japan's major cities. In 1967, the Tokyo metropolitan government limited public sector jobs in the city's major packinghouse in Shibaura to sixty. *Buraku* performed additional work necessary to slaughter and process the plant's cattle and hogs without direct pay. That is, the metropolitan government said that if *Buraku* wanted to continue to buy edible by-products from the facility then they should provide labor to help with the slaughtering process. This so-called "labor without pay" system resulted in the formation of the All Shibaura Slaughterhouse Employee Association established in December 1971. The employee association went out on strike at the Shibaura plant in August 1973, which then helped to establish a BLL branch in Shibaura. From 1974 to 1980, the employee association and BLL branch worked together in several strikes, sit-ins, and other political actions to attempt to force the metropolitan government to pay all workers for work in the slaughterhouse and improve the status of the public employees who worked in the meat industry. BLL support for packinghouse workers in both Ōsaka's and Yokohama's public slaughterhouses was also crucial in the establishment of employee associations and successful recognition of *Buraku* rights as public employees. Unionization and government recognition of workers' rights took place from 1967 to 1970 in Ōsaka and in 1979 in Yokohama.[19]

State support for the union-building movement in the US meatpacking industry was crucial in the 1930s and 1940s not only in improving workers' workplace rights and economic benefits, but also in helping the labor movement incorporate large numbers of African Americans into the industry. This support was crucial given the long history of ethnic and racial tensions among meat workers. African Americans were recruited into US meatpacking plants in large numbers initially as strikebreakers, especially during the American Federation of Labor's Amalgamated Meat Cutters and Butcher Workmen's nationwide strike in 1904. In Chicago alone, an estimated 10,000 to 18,000 African Americans were brought into the stockyard district to take the place of about 23,000 workers who participated in the strike. Among large numbers of white AFL union members, hatred of strikebreakers, particularly black strikebreakers, was endemic.[20] One AFL leader in 1904 said black strikebreakers were "huge, strapping fellows, ignorant and vicious, whose predominating trait was animalism."[21] For most African Americans, however, strikebreaking was not a matter of siding with employers against unions but simply a matter of pursuing gainful employment in an environment in which blacks were largely excluded from such opportunities. Nevertheless, white union members' anger against black strikebreakers in the 1904 meatpacking strike reinforced an image of blacks not only as ignorant and vicious but also as incapable of humane treatment of animals.[22]

In Chicago, union portrayals of black strikebreakers of the 1904 strike emphasized their general immorality, playing up newspaper reports of gambling, drinking, and sexual debauchery in the packinghouses. Press accounts emphasized how the stockyards district had become a place for shooting craps, poker games, and even nightly prizefights. The packing companies housed the strikebreakers inside the plants and surrounded them with armed guards, which, of course, allowed white imaginations to run wild with all manner of possible black depravity. Union accounts also emphasized how blacks were incompetent butchers. The Amalgamated Meat Cutters and Butcher Workmen's president, Michael Donnelly, claimed that in East St. Louis "cattle are being assassinated, not butchered." Instead of stunning cattle with one well-placed blow, Donnelly claimed that it often took black men eight blows and that animals were often left partially stunned on the kill floor. Instead of pursuing humane slaughter, Donnelly said it was a "bull fight on the killing floor."[23]

Despite white workers' mostly hostile attitudes toward them, blacks became an increasingly numerous component of the American meatpacking industry's workforce through the first third of the twentieth century. This was especially true in Chicago but was also the case in many other cities with major stockyards facilities in the Midwest. The Congress of Industrial Organizations (CIO) union movement in Chicago and elsewhere across the Midwest during the 1930s to 1950s did much to dissipate the perception that African Americans were incompetent butchers and, accordingly, illegitimate participants in the meat-making process. The CIO's national leadership saw the federal government's protections for union-building and certification elections as a new opportunity to build mass-production industrial unions that could bring economic and workplace benefits to the growing number of unskilled and semiskilled workers, including African Americans. The CIO encouraged black workers to take on leadership positions in the Packinghouse Workers Organizing Committee beginning in 1937 and later in its successor, United Packinghouse Workers of America (UPWA), in 1943. As early as 1939, nine of the fifteen meatpacking local unions in Chicago had black presidents. The union became much more racially united, not only in leadership roles but also in terms of wages. By 1940, blacks' wages in the meat industry were almost equal to whites, making it the most successful industry in this regard of any in the North.[24]

In Western Europe, state support for public slaughterhouses often implicitly signaled support for butchers' control over the slaughtering processes, a situation that was not true in the United States, where butcher authority was steadily eroded even in union workplaces. In France, public slaughterhouses were primarily meant to ensure stable prices for sufficient fresh meat. The system, however, did not interfere with the traditional craft methods of the wholesale butchers who worked in these packinghouses, particularly La Villette, the primary public slaughterhouse in Paris after 1867. In contrast to the American private slaughterhouses, where employers consistently eroded butchers' craft skills and modified their methods of making meat, butchers retained their craft autonomy in the public slaughterhouses in France into the twentieth century.[25]

The story in Britain was more complicated. There, government officials and humanitarian reformers challenged the craft control of butchers in the private sector. Public slaughterhouses in Britain in the late nineteenth century became associated with imports of chilled and frozen beef and other meat from the Americas, New Zealand, and Australia. Private slaughterhouses in London declined from 692 in 1889 to 411 in 1899 to 273 in 1908 before being abolished altogether in 1927; but many dispersed to other parts of the country and continued to provide meat for sale. For the most part, private butchers resisted many of the industry changes associated with food safety, environmental reform, and humane animal slaughter because the butchers saw these changes as challenges to their traditional craft authority.[26]

In both Australia and New Zealand, state marketing boards were established to promote the two countries' meat exports, particularly frozen mutton, lamb, and beef for the British domestic market. The Australian Meat Council was founded in 1925 followed by the Australian Meat Board in 1936, while the New Zealand Meat Producers' Board started in 1922. The boards helped to increase meat exports to Britain through the 1920s and then stabilize the two countries' roles as suppliers to the British market at the Imperial Conference in Ottawa, Canada, in 1932. The producer boards also helped to stabilize family-farming-based livestock production as well as labor relations in the meat industries in both countries through the 1970s. The Australasian Meat Industry Employees Union had a long history of works-council-based representation. New Zealand's Industrial Conciliation and Arbitration Act passed in 1894, which remained in place for another ninety years, encouraged labor union formation and provided compulsory arbitration between employers and employees. The New Zealand Meat Workers' Union and various regional Freezing Workers' Unions benefited increasing numbers of Māori workers, as well as Pacific Islanders who were drawn to industrial work, especially during the 1950s and 1960s. Māori men and some women continued to gain employment in the freezing works during the 1960s and 1970s. By the 1970s, Māori made up the majority of the North Island's freezing workers, especially in and around Auckland. Although the industry remained largely seasonal, the Auckland Freezing Workers won equal pay for women in 1971, paid sick leave in 1972, and wage gains.[27]

## The Corporate Food Regime and Increasing Animosity toward Meat Workers

Around the world beginning especially in the 1980s, TNCs, such as ConAgra, Cargill, and Tyson, pursued global sourcing strategies that helped to undermine state support for industrial relations built on labor unions and workplace protections. In the global meat industry, as economic benefits and workplace conditions eroded, corporations pursued labor recruitment strategies that exacerbated ethnic and racial conflicts both among employees and by the public toward meat employees. In addition to the steady deterioration in workplace and economic benefits, because of the resurgence in public concerns about humane slaughter and animal protections, TNCs, as well as their employees, face increasing hostility about animal abuse. In a context of weakening government protections and declining union power, differences among greater numbers of marginalized immigrant meat workers in terms of language and culture have further weakened workers' abilities to respond to such charges and create public awareness of their struggles.[28]

The post–World War II evolution of the meat industry in the United States illustrates the range of problems that workers now face. Starting in the 1950s, the American meatpacking industry underwent terrific economic changes. The older corporate giants in the industry were either purchased or supplanted by new giants, especially Iowa Beef Packers (later, IBP, which was then purchased by Tyson Foods), ConAgra/Swift, and Excel/Cargill. These new companies revolutionized the economics of the industry. One of these major economic changes was a shift to production in rural and more isolated Midwestern communities. The CIO-led union-building efforts that had created greater racial harmony disintegrated when the older plants were closed. As the new meat corporations added plants, they also increasingly hired Latino, Southeast Asian, and even African workers. Between 1980 and 2005, two of every five individuals who moved to the Midwest was Latino. By 2015, Latino workers made up almost 40 percent of all meatpacking employees. The South, due largely to the growth of the poultry industry there after World War II, also attracted greater numbers of Latinos starting in the late 1980s. Poultry company recruitment strategies as well as the 1986 Immigration Reform and Control Act facilitated this transition. Although the 1986 act was intended to help undocumented immigrants (by allowing unauthorized aliens to apply for legal status) and to prohibit companies from knowingly hiring undocumented immigrants, it actually encouraged the proliferation of undocumented workers available to industries, including meatpacking. Between 1990 and 2000, for instance, Arkansas's Latino population increased by 337 percent. By the 1990s, older residents of Nebraska often referred to their state's meatpacking communities as "Little Mexicos."[29]

The new meat companies recruited immigrant labor primarily because they could pay them lower wages. As companies closed or threatened to close plants, they forced the major packing union—the Amalgamated Meat Cutters and Butcher Workmen (AMCBW), AFL-CIO, formed in 1968 from the merger of the UPWA-CIO and the original AMCBW-AFL—to make significant concessions. As a result, packing wages fell by 50 percent from the mid-1970s to mid-1980s. By 2013, meatpacking workers' real wages were 40 percent less than they had been forty years earlier. As closures and restructuring continued, the AMCBW merged with the Retail Clerks International Union in 1979 to become the United Food and Commercial Workers (UFCW). Union contracts in the industry fell by a third from 1976 to 1983, and more plants were opened as nonunion workplaces. In this context, workplace conditions eroded in the face of increased division of labor and faster line speeds. In 1991, meatpacking workers were three times more likely to be hurt on the job than the average US manufacturing worker. In addition, the new immigrant meat workers faced not only poor wages and deteriorating workplace conditions but also substandard housing and lack of access to bilingual education and other social services. Despite the fact that meat corporations actively recruited undocumented workers, immigrants faced periodic raids from the United States Immigration and Customs Enforcement and community ordinances to bar them from working or renting inside city limits. While some communities attempted to provide assistance for newcomers' social, cultural, and educational needs, the process of transplanting people from the far corners of the world inevitably led to chaotic, divisive, and prejudicial views toward the workers. The ghettoization of plant workers exacerbated all of these feelings. Moreover, immigrants moved to small midwestern and southern towns where they competed for jobs with whites and blacks who were often displaced from the collapsing rural economy. All these factors created hostility to the newcomers that has yet to dissipate.[30]

In New Zealand, the global reorganization of the meat industry resulted in similar erosion of union protections among meat workers, which disproportionately affected the Māori and Pacific Island workers in the industry. Many of the country's farmer cooperative and locally owned meat companies were bought out by corporations, including Britain's Vestey Group. During the 1980s, the ruling Labour Party supported the largely export-driven meat industry's efforts to deregulate and eliminate alleged production inefficiencies through wage cuts and plant closures. A major meat industry strike in 1986 resulted in divisions between the Meat Workers' Union and the export-oriented freezing workers' unions. Between 1986 and 1990, the freezing workers' unions lost over half of their membership in the industry shake-up. Indeed, between March 1987 and March 1989, one-fifth of the Māori work force lost employment. At the same time, the country's two dominant union groups, the Federation of Labour and the Combined States Unions, which merged in 1987 to become the Council of Trade Unions, largely rebuffed Māori initiatives to recognize Māori self-determination, representation of all union decision-making bodies, and acknowledgment of Māori sovereignty efforts.[31]

The use of increasingly marginalized immigrant meatpacking workers in the 1980s coincided with the rise of the modern animal rights movement and animal liberation ethics. Although it is difficult to unravel prejudices against immigrant workers and the perception that meat workers in large packing plants engage in animal abuse, many Americans believe this to be true. In one of the more sensationalist accounts of the modern industry's horrors, Gail Eisnitz's *Slaughterhouse* contains firsthand accounts of animals tortured by careless and inhumane modern meatpacking killing practices, side-by-side with depictions of the abusive treatment of the industry's largely immigrant workforce. Given how she conveys the many abuses associated with modern meat production, accounts of animal abuse tend to run together with depictions of the immigrants who work in the plants.[32]

Although Eisnitz did not attribute animal abuse to particular ethnic groups among the workers, it is difficult for her reading audience not to make that association. This connection is also suggested in various government reports. For instance, a 2004 US Government Accounting Office report found that hundreds of thousands of animals were not properly stunned before being slaughtered, resulting in fully conscious animals being killed. A May 2013 USDA Office of the Inspector General report found that eight of thirty hog slaughterhouses had seriously violated the Humane Slaughter Act, passed in 1978. Workers' injuries have also increased in recent years as Hazard Analysis and Critical Control

Points–Based Inspection Model Project methods (HIMP) have been increasingly implemented since the late 1990s. This method gives greater control over line speeds to meat companies and decreases USDA inspection oversight. The 2013 USDA report linked many of the animal abuse incidents to use of HIMP.[33] Timothy Pachirat's recent book on the elaborate division of labor in modern beef slaughtering plants that allows one steer to be killed every twelve seconds emphasizes that such labor is "considered morally and physically repellent by the vast majority of society."[34] Despite efforts to reform some elements of modern meatpacking working conditions, meat workers who are employed for five years in a plant today have a nearly fifty-fifty risk of suffering a serious injury. Moreover, as immigrants became increasingly numerous across the Midwest's and South's meatpacking communities, they could not help but be associated with the number of food safety violations that continually plague the industry.

Increasingly during the late twentieth and early twenty-first centuries, TNCs have pursued similar strategies as those that operate in the United States and New Zealand. Plants located in the developed world have relied on ever increasing numbers of migrant workers from the less developed regions of the world. In the United Kingdom, where nearly 40 percent of workers of the nearly ninety thousand employees in the meat industry were estimated to have been migrants in 2013, reports of abuse of foreign workers are constant.[35] In one report published in *The Guardian* in 2010, a fifteen-month inquiry found that among the foreign workers—mostly Polish but also some Lithuanian, Latvian, Czech, Slovakian, and Portuguese employees—20 percent reported that supervisory personnel had physically abused them. A Brazilian man who worked in an east England poultry plant said, "I'm not a slave. I just can't speak English. [My supervisor] talked to me like he talked with an animal."[36] A 2011 report on the Kepak company in Ireland, one of the leading meat processing firms there and a contributor to Ireland's leading status as a beef exporter to the European Union (EU), noted that foreigners, especially Poles and Latvians but also Brazilians and Chinese, made up a majority of the company's workers. In Germany between 2001 and 2014, the meat industry increasingly hired Romanian and Bulgarian workers to work night shifts as part of a trend to replace unionized German workers with nonunion subcontractors. The trend in Australia has also been toward increasing use of foreign workers, including migrants from Mexico, Uruguay, Brazil, Singapore, Taiwan, Malaysia, and various Pacific Islands. A 2014 report estimated that of the more than fifty thousand workers in Australia's meat industry about ten thousand were temporary international workers. Many of these workers were underpaid, mistreated, and threatened with dismissal if they complained about their working conditions. In China, where Buddhist traditions frowned on cattle slaughter, much of the cattle slaughter is carried out by people of Hui ethnicity, who are typically Muslim.[37]

In Japan, despite significant post–World War II government intervention in alleviating some of the traditional hardships faced by the *Buraku* population, the public still exhibits prejudices against them. In 2003, Miura Koukichirou outlined some of the ramifications of such prejudices. Because the vast majority of Japan's slaughterhouses were small—normally ten to twenty employees, who were typically of *Buraku* heritage—and based in urban neighborhoods, many of the movements against animal slaughtering were advocated under the guise of environmental and educational concerns. Miura explained that neighborhood movements in Fukushima and Nara, for example, claimed that slaughterhouses created water pollution and violated government guidelines prohibiting placement of such facilities close to schools. Miura countered that such excuses were longstanding in the history of *Buraku* prejudices; Japan's modern history is rife with cases of slaughterhouses forced to move because of pollution claims. In fact, according to Miura the problem with water pollution was typically due to municipalities refusing to pay the cost of updated water purification systems. In one case, a municipality refused to provide a restroom for a plant's workers. Miura also pointed out a need for Japanese schools to provide education on the purpose of slaughterhouses to help root out the underlying longstanding cultural prejudices among Japanese people toward such work and its association with *Buraku*. In Miura's view, this underlying discrimination motivated municipalities to urge consolidation of smaller slaughterhouses into larger public facilities that could be removed from traditional urban locations. Miura claimed that about 270 of Japan's 300 slaughterhouses were small to medium sized. In a similar fashion the series of international trade agreements that had opened Japan's meat markets to foreign imports beginning especially in the 1980s were eliminating *Buraku* from longstanding employment that, no matter how despised by the larger public, had also been stable and fairly lucrative.[38]

In 2004, Sakurai Atsushi published an account of the conditions for workers in a small slaughterhouse in Oumihachiman, Shiga Prefecture. Located in a part of Japan with longstanding associations with meat and leather production, there had been two slaughterhouses in Oumihachiman until the 1970s. Both were publicly operated by the village government beginning in 1956. Similar to the situation described by Miura, most of the butchers and leather workers were *Buraku*. Some indicated to Sakurai that no one had ever looked down upon them, while others said that they had experienced prejudices. Perhaps most telling about townspeople's attitudes was the fact that the plant manager would not allow Sakurai to photograph the faces of any of the workers for fear of the "reputations of the men." Neither would the leather workers allow their photographs to be taken for fear of repercussions against their children.[39]

While meat workers share many of the same problems faced around the world by food workers, the particular animosities generated by their involvement in killing huge numbers of animals have made them a target for scorn in ways not shared by other food workers. State intervention over much of the twentieth century improved their status and public perceptions, but the combination of diminishing government safeguards and rising concerns over animal ethics and protections since the late twentieth century has once again made them the targets of hostility. Henry Salt's characterization of meat workers as a "pariah class" is especially notable in light of new emphases on locally sourced, artisanal produced meat. As noted in chapter 9, many affluent Western meat consumers motivated by concerns about animal ethics have developed a new version of Henry Bergh's "necessary killing" in choosing to buy meat from small producers rather than meat produced by a large corporation using immigrant labor. In the United States, for example, where small-scale livestock production accounts for only about 1 percent of total meat production, the growing popularity of farmers' markets offers consumers the option to buy meat from small producers, who they assume neither abuse their animals nor exploit those involved in helping them slaughter their animals. Between 1994 and 2014, the number of farmers' markets jumped from 1,755 to 8,284. However, only a relative handful of small producers actually pursue certification of humane treatment. While affluent consumers seem increasingly alarmed about meat workers' treatment and possible connections to animal abuse, their primary concerns focus on meat's connections to environmental and public health problems, the subjects of the next two chapters.[40]

# 8

## A Plague of Pigs and Other Environmental Dilemmas

MEAT'S PROLIFERATION IN PEOPLE'S DIETS SINCE World War II has created or exacerbated many human environmental effects. Since its origins several thousand years ago, agriculture's very nature has fundamentally transformed the environment. Intensification of the livestock and meat revolutions that have been underway around the world over the past sixty years has made environmental concerns more pressing than at any previous time in human history. The effects of CAFOs and large-scale meat production are many and complicated, including enormous land uses, especially when the necessary feed crops are factored into the calculations; pesticide and toxic chemical applications; huge water needs and water pollution; and significant greenhouse gas emissions, particularly carbon dioxide, methane, hydrogen sulfide, and nitrous oxide. This chapter focuses on intensive livestock and meat production's impact on land, water, and air around the globe that highlight how the post–World War II growth in meat production and consumption has created new problems.

Around the world today, intensive livestock production has become a more vexing source of environmental degradation than production of meat itself. CAFOs are typically defined as production sites with 1,000 or more animal units—that is, 1,000 cattle, 2,500 pigs over 55 pounds, 55,000 turkeys, or 125,000 chickens (broilers)—where animals are confined for at least 45 days during a growing season in an area that does not produce vegetation. Among the multitude of possible global environmental impacts that might be discussed here—including increasing feral pig populations, livestock's contributions to declining and fragmenting ecosystems, soil degradation, and pesticide and pharmaceutical effects—this chapter examines two areas of concern: issues tied to the increasing amounts of livestock manure and wastewater and cattle's contributions to atmospheric concerns about odors and greenhouse gases.[1]

## Livestock Manure and Wastewater Pollution

Livestock manure has traditionally been one of the most valuable by-products of animal production. Depending on the time and place, manure has sometimes been even more valuable than the meat produced in animal slaughter. After chemical fertilizers began to be developed and used in the West, studies of manure's use in China, Japan, Korea, and India in the early twentieth century demonstrated that traditional organics were as or even more effective than inorganics.[2] Franklin Hiram King, a former agriculture professor at the University of Wisconsin, found in his trip to East Asia in 1909 that both animal and human waste was being used in "marvelous account in the maintenance of soil fertility and in the production of food."[3] He watched people in China search for and collect animal droppings along country and caravan roads. Sir Albert Howard, an English botanist, used his studies of crop production in India in the early twentieth century to develop the Indore Method of compost production, which used all available vegetable matter, including soiled cattle bedding, crop residues, fall leaves, forest wastes, animal manure, green (i.e., legume-based) manures, weeds, urine earth from cattle-shed floors, and wood ashes. Even today in the United States and Europe, most of the manure that livestock generate is still applied to crop and pasture land. Nevertheless, because of their number and scale, intensive CAFO livestock production systems—starting with the poultry industry during the 1940s, followed by the hog industry combined with cattle feedlot systems in the generation following World War II—have transformed manure into one of the most dangerous environmental threats associated with livestock production throughout much of the meat-eating world. By 2000, nearly 80 percent of the world's total pork, poultry, and egg production occurred in CAFOs.[4]

Until the 1960s in the United States and much of the rest of the world, little attention was paid to livestock production's environmental effects. Considerable attention, however, was focused on water and air pollution caused by meat production. These concerns accelerated during the nineteenth century, as meat production steadily increased in Europe and North America, resulting in greater municipal government oversight over meatpacking's location and the wastes produced by such facilities. Court battles between meatpackers and city governments were frequent as well. In New York City, Chicago, and St. Louis, legal measures attempted to contain pollution primarily by punishing slaughterers until they developed by-product industries instead of dumping blood and waste in waterways. Similar measures passed outside of Europe and North America. Australia passed a Noxious Trades and Cattle Slaughtering Act in 1894 to allow municipal governments' boards of health to regulate slaughtering and rendering. In Japan, where meat production had traditionally been confined to *Buraku* areas of cities, the Noxious Fumes Prevention Act of 1971 focused on water and air pollution near urban meatpacking plant locations.[5]

In the United States between the First and Second World Wars, numerous meatpacking plants embarked on construction of sewage and wastewater facilities that could treat the water, blood, grease, and fecal matter produced in the process of making meat. Plants located in both large and medium-sized urban locations constructed such facilities and effectively removed much of the sludge that had previously been discharged into nearby rivers. Indeed, many of these systems recovered sludge and processed it to retain protein, nitrogen, and ammonia. Most of the pre–World War II treatment facilities used aerobic processes, such as activated sludge removal. The channel-aeration activated-sludge (or Pasveer) process, which had been used extensively in Western Europe in the 1950s and 1960s, was incorporated into more than one hundred packing plants' treatment systems in North America in the 1960s and 1970s. These new systems required more extensive pretreatment, especially in terms of grease and solids removal, before rotor and floating aerators combined the sludge with chlorine to treat and remove wastes and fecal matter. The treated sludge could then be transformed into animal-feed supplements. Nevertheless, water pollution from packing plants still occurred. Millions of dollars of fines were levied against packing plants in the United States during the 1980s through early 2000s for failure to adequately treat meatpacking waste.[6]

Before the number and scale of CAFOs accelerated after World War II, however, pollution from rural areas was of little concern. In the United States, the Water Pollution Control Act (PL80–845), passed in

1948, established federal responsibility for controlling pollution in interstate waters. However, states assumed responsibility for enforcement of CAFO pollution regulation under the Clean Water Act (PL92–500) of 1972; and states and local governments had jurisdiction over manure disposal regulation. Generally speaking, smaller scale livestock production results in nonpoint source pollution, which is more diffuse and not effectively addressed by the Clean Water Act. As CAFOs became more prominent, they generated point-source pollution and, ostensibly, could be better controlled and regulated.[7]

As is true of meatpacking waste products, anaerobic or aerobic processes and sometimes combinations of both can be used to treat livestock confinement waste. Different types of animals produce waste with different physical and chemical properties, a fact that further complicates issues related to most effective treatments. Animal manure contains nitrogen and phosphorus, both of which are quite valuable in limited quantities but in large quantities cause damage to waterways and the atmosphere. Nitrates in drinking water can cause human health problems and contribute to eutrophication and oxygen-depleting algae blooms; when converted from organic to inorganic forms, nitrogen forms ammonia, producing odor problems and acid rain. Excess phosphorous, like nitrates, in waterways also contributes to eutrophication. In CAFO environments, chickens and turkeys produce litter, which is manure mixed with bedding material. Traditionally, litter was used as fertilizer. Pigs typically produce waste that is captured and treated in either lagoons or pits. Cattle and dairy cow waste is usually handled as a slurry or kept dry. In terms of nitrogen, chickens and turkeys produce the most concentrated type, followed by swine and then cattle. With regard to phosphorus, swine manure has the highest concentrations. Chicken manure usually has the highest concentrations of hormones. Decomposition of manure results in carbon dioxide, methane, ammonia, and hydrogen sulfide, all of which can kill both humans and animals. The fecal coliform bacteria and other parasites present in animal manure can cause a variety of zoonotic diseases.[8]

In the United States and Europe, and increasingly in China, many of the most troublesome issues of animal manure waste disposal have involved CAFO hog production. Around the globe, hog production has intensified immensely in the past two decades. In the United States between 1998 and 2009, the total number of hog operations fell by 60 percent, but the average inventory on the remaining farms increased threefold, from 2,589 to 7,930 head. Hog density, as measured by the number of animals per square mile, increased even more rapidly in the Netherlands, Taiwan, Denmark, South Korea, Japan, Germany, and Spain. In fact, of the two major hog-producing states in the United States since the 1990s—Iowa and North Carolina—only North Carolina had hog densities as high as any of these countries in 2000. Two major types of hog manure treatment systems are currently used: lagoon and pit. Anaerobic lagoons were first used to treat livestock waste in the mid-1960s. Lagoon systems, used in the United States, particularly in North Carolina, reduce the nitrogen content of the hog manure through anaerobic digestion and ammonia volatilization. The latter creates problems in terms of atmospheric pollution and acid rain. Pit systems, used especially in Iowa, capture the manure in tanks or pits located directly underneath the confinement facilities. Manure falls through slots in the floor and is then treated. In both cases, though more so in the case of pit systems, after treatment the manure is generally spread on land, either on site or off site, as fertilizer. The percentage of pigs in the United States raised in pit versus lagoon systems increased in the decade from 1998 to 2009. In cases where lagoons are covered, as the next section expands upon, the biogas can be captured, recovered, and used to generate energy, typically to power the CAFO facilities. Such systems are much more common in Europe and Asia than in North America. Europe and Asia also have more treatment systems that are able to work like municipal sewage systems and allow the effluent to be cleaned adequately for direct flushing into waterways.[9]

Several environmental disasters associated with hog manure occurred worldwide during the 1990s. In the United States, North Carolina was the location of many of these occurrences, particularly along its eastern coast. Between 1989 and 1995, the state's hog population increased by 85 percent, which led to a jump in North Carolina's rank among the US hog-producing states from fifteenth to second. Smithfield Foods, which also become the nation's largest pork producer during the 1990s, owned about three-quarters of the state's hog CAFOs. At the beginning of the 1990s, hog CAFOs were treated as farms for zoning and environmental regulation purposes. While the explosion of production rapidly harmed the coastal fisheries and crabberies, it was not until hurricane damage caused several lagoon spills culminating in the 25-million-gallon lagoon discharge into the New River in 1995 that groups mobilized to limit environmental damage. The environmental group Waterkeeper Alliance and one of its attorneys, Nicolette Hahn Niman, worked out an agreement with Smithfield's subsidiary, Murphy-Brown LLC, to take steps to limit production. The state then passed a Clean Water Responsibility Act in 1997, which issued a ten-year moratorium on the building of new hog CAFOs and waste lagoons. Indeed, this is one of the main reasons why pit-treatment systems began to outpace lagoon-treatment systems nationwide during the early 2000s. While Niman placed much of the blame on Smithfield, at least one study showed that hog producers in the state were aware of environmental threats associated with manure lagoons, but only an estimated one-third of them adopted appropriate environmental management practices.[10]

In Europe's and Asia's major hog-producing countries, such as the Netherlands, Denmark, Germany, Spain, and Taiwan, hog manure's effects on water resources are also an environmental concern. For instance, in Spain, which was Europe's second-largest hog producer at the turn of the twenty-first century, pig manure is considered the country's most important externality associated with its intensified hog production. Much of this production occurs in Catalonia, which has particular hydrogeological conditions that make the seepage of nitrates from the manure especially damaging to ground and surface water. Efforts to reduce these problems by spreading more manure on farmland and to use hog manure in energy cogeneration plants have only partially addressed the issues. Beginning in the 1980s, Denmark required its hog producers to store liquid manure in impermeable tanks. The Netherlands passed a Manure and Fertilizer Act in 1987 to prevent increases in manure production. Several EU countries in the 1990s banned manure applications during certain months. Denmark also started building biogas plants (anaerobic digesters) in the late 1980s and had twenty in operation by the early 2000s. Japan requires all hog producers to develop manure management plans. Taiwan treats its hog manure in systems akin to municipal sewage treatment plants so flushed water can be discharged into surface waters or recycled as potable water on hog production sites. Mainland China, which produced 15 percent of its hogs in the 1990s in CAFOs compared to 75 percent by 2005, faces enormous water pollution issues, particularly since it produces one-half the world's pork and the Chinese Ministry of Agriculture believes that it can best maintain quality food production standards by integrating its CAFOs with its processing plants.[11]

A variety of mitigating environmental practices and stiffer governmental interventions in the United States and Europe regarding hog manure management were put into place since the 1990s. Following practices that various European nations had started to use in the 1950s, US hog producers began to use wetlands, both natural and constructed, to help treat wastewater. Better results are generally attained through designs of engineered wastelands. Particularly in North Carolina and other southern states with large-scale hog manure management concerns, Bermuda grass is grown to help consume and dispose of hog waste. Composting has also increased in both the United States and Europe, though the Dutch government in the 1990s had difficulty trying to market the composted material. Numerous states in the United States passed CAFO regulations in the 1990s and early 2000s. Even with these new efforts underway, the United States Environmental Protection Agency (EPA) reported in 1998 that livestock waste had polluted 35,000 miles of rivers in 22 states and polluted ground water in 17 states. The US Clean Water Act was revised in 2003 and new regulations implemented in 2008 to better control runoff from the largest CAFOs, which by 2010 were estimated to produce a total of more than 500 million tons of waste each year. The new regulations require CAFOs that discharge or propose to discharge manure effluent to file permit coverage with the National Pollutant Discharge Elimination System (NPDES). Similar filing and reporting programs were created for atmospheric pollution, as explained in the next section. The USDA's program to help smaller livestock producers pay for necessary environmental protection measures, the Environmental Quality Incentive Program (EQIP), which began in 1996, was expanded in 2003 to allow large CAFOs to apply for financial assistance. Over the span of the early 2000s, the total number of farms receiving EQIP payments increased from 1.5 to 4.4 percent. While important in terms of overall environmental impact, US taxpayers are helping to defray CAFO environmental cleanup expenses. The EU also continued to move even more aggressively toward environmental cleanup measures but at great expense to European taxpayers.[12]

While hog manure has received the most attention in terms of its role in creating water pollution problems, cattle and poultry manure also raise multiple concerns. As explained in the next section, much of the concern with both beef and dairy cattle is their role in producing greenhouse gases. In the United States, beef cattle manure produced on feedlots is almost always disposed of through application to pasture, but the EPA has found that nearly all feedlots have insufficient land for safely disposing of such waste. Beginning in the 1940s, when confinement of poultry began in the United States, studies were conducted to see if litter could be fed to cattle. Ruminant animals, including cattle, can digest both non-protein nitrogen and the plant fibers found in bedding. Of course because the litter contains a wide variety of bacteria, the litter must be processed first to rid it of disease. Converting litter to feed pellets has been suggested as one way to cope with and profit from poultry manure. In the United Kingdom, some broiler litter has been used for on-farm space heating of the chicken houses as well as off-site electricity generation.[13]

## Livestock's Air Quality Effects: Odors and Greenhouse Gases

Concerns about odors are both subjective and objective and, in the Western world, also clearly tied to the fact that meatpacking moved from urban to rural or suburban locations. Whereas urban residents may have been used to a variety of industrial and other unpleasant odors, suburban residents more often are not and have not felt as kindly disposed to ignore them. When animal slaughtering was moved from urban to rural locations throughout Western Europe and North America during the late nineteenth century through the mid-twentieth century, concerns about odors were part of the debate about the merits of these relocation decisions. The location of livestock operations has also been contentious in terms of odors, even when animals are ostensibly situated in rural areas. Part of this is because of the size of animal confinements. Thousands of animals in concentrated confinement or feedlots create odors that are difficult to mask even when located many miles from most people's residences. Careful engineering of wastewater lagoons and even additions of mint oil to animal feed might mask the odor of cattle urine. Starting especially during the 1980s, residents located near livestock containments throughout the US's Midwest, East Coast, and rural South increasingly pushed for local, state, and federal intervention to help abate odor problems. Considerable research in the United States and Europe has indicated that livestock odors, which consist in part of ammonia, hydrogen sulfide, and other volatile organic compounds, can result in lung diseases, nausea, nosebleeds, depression, and even brain damage, and are also hazardous to the livestock themselves. Some livestock waste experts, such as J. Ronald Miner, have asserted that odors are the greatest challenge among all the airborne pollutants associated with livestock production. A 2003 University of Iowa study found that factory farm workers were exposed to 331 volatile organic compounds and gases in the air that they breathed.[14]

Although livestock odors are a significant concern partly because they are so obvious and affect the quality of human and animal life, as an atmospheric pollution issue, livestock's, particularly cattle's, contribution to greenhouse gases—those that absorb and emit infrared radiation—is a much more daunting global concern today. Indeed, Simon Fairlie maintains that since 2006, when the FAO issued its report titled *Livestock's Long Shadow*, the major argument against meat eating has been cattle's contribution of methane to global warming. According to the report's calculations—which are used in virtually all studies that reference the topic of livestock and greenhouse gases—total anthropogenic, or human-made, contributions from livestock of methane production comprise 37 percent. About 60 percent of overall methane emissions come from natural sources, particularly wetlands. Methane is calculated to be twenty-five times more powerful in global warming potential over a one-hundred-year life span than carbon dioxide. That is, each gram of methane has the same global warming potential as twenty-five grams of carbon dioxide. Cattle's production of methane is a particularly vexing problem because ruminants, which include sheep and goats as well, naturally produce methane as a by-product of their digestive systems.[15] In Fairlie's words, "[m]ethane is integral to the metabolism of the ruminant, and indeed to the entire biological world."[16] This enteric, or intestinal digestive, production of methane accounts for cattle's and other ruminants' greatest direct contribution to methane. Approximately 92 to 98 percent of the enteric production is released from the cattle and other ruminants' mouths as belches, while the remaining 2–8 percent is released as flatulence. Based on calculations in *Livestock's Long Shadow*, about 80 percent of cattle and other ruminants' methane come from enteric production, with the other 20 percent from other sources, such as production of livestock fuel and feed through synthetic nitrogen fertilizer production and organic decomposition that includes animal waste products.[17]

Livestock's other major contribution to greenhouse gas production is nitrous oxide ($N_2O$), which is estimated to be at least two to three hundred times more powerful as a greenhouse gas than carbon dioxide. Globally, agricultural activities are the largest source of nitrous oxide emissions. Livestock activities account for about 65 percent of these sources. About 60 percent of livestock's contribution of nitrous oxide, according to *Livestock's Long Shadow*, comes from manure emissions, manure management, and manure application. Nitrous oxide results from nitrification and denitrification—biological processes that respectively oxidize ammonia to nitric acid, nitrous acid, or any nitrate or nitrite and reduce nitrates or nitrites to nitrogen-containing gases. Partly because the science of livestock and greenhouse gases calculations is so complicated, researchers note a lack of reporting and public understanding about the contribution to greenhouse gases from livestock and agricultural activities more broadly. Most commentators on the topic of anthropogenic sources of greenhouse gases focus on auto and industrial emissions, not agriculture, including production of synthetic fertilizers or livestock.[18]

Concerns about the connection between cattle and methane became prominent during the 1990s, particularly when the expansion of cattle-raising in the Brazilian Amazon accelerated. Beginning during the mid-1970s, Amazon smallholder farmers increasingly adopted cattle ranching as a more lucrative agricultural endeavor. Over the decade of the 1990s, about 60 percent of global deforestation occurred in the Amazon, as Brazil's absolute cattle numbers, particularly those raised in deforested areas, continued to grow. Of course, larger numbers of cattle produced more methane, but just as important, the increasingly denuded Amazon removed trees from their role in absorbing carbon dioxide from the atmosphere. As Fairlie points out, the impact from carbon dioxide on greenhouse gas production may be more important than methane's since the latter converts to less harmful chemicals within twelve years whereas carbon dioxide has an atmospheric lifetime of two hundred years. Because of worldwide outcries and political pressures, the Brazilian government took steps in the 2000s to slow deforestation. In 2009, four of Brazil's largest beef companies agreed to a moratorium on buying cattle from newly deforested land. Rates of Amazon deforestation slowed considerably during the 2000s, though commentators like Fairlie point out that the clearing of Brazil's Amazon is likely tied more closely to prices of soybeans and the country's expansion of that crop's production.[19]

Although most developed nations' livestock sectors have been slow to acknowledge their role in producing greenhouse gases, several countries starting in the 1990s took steps to both calculate their contributions and mitigate the effects. In 1994, Australia's meat and livestock industry published a report that calculated that agriculture overall contributed about 30 percent of the nation's total greenhouse gases while livestock's total contribution of greenhouse gases was 8 percent. Livestock production created 62 percent of the country's total methane emissions while livestock urine and manure constituted three-quarters of the nation's nitrous oxide emissions. The report noted, however, that Australia's methane emissions comprised just 0.32 percent of the world's total. Although the report noted that the cost of reducing its emissions through shrinking the livestock industry itself would be prohibitively high, measures were noted about how enteric methane might be reduced through modification of cattle and sheep intestinal microorganisms, by modifying feed supplements, and through more careful manure management practices. Another report issued in 2001 noted that agriculture's contribution to greenhouse gases had declined to 20 percent of the total. Methane production had declined between 1990 and 1998 because of a large reduction in the nation's sheep industry. Methane emissions had declined overall by 5 percent during that period. According to this report, many of the recommendations from the 1994 report had been implemented. Likewise, New Zealand's meat and livestock industry was reluctant to accept a role in the production of greenhouse gases, but the country's Emissions Trading Scheme, established in 2008, included agriculture in total greenhouse gas emissions with plans to develop ways to mitigate its contributions to this problem.[20]

EU nations have also made some efforts toward identifying and mitigating livestock production's effects on the atmosphere. The EU adopted a thematic strategy on air pollution and decreased agricultural emissions overall by 20 percent between 1990 and 2005. Both the Netherlands and Denmark, which have fairly stringent laws pertaining to livestock effects on water quality, require careful monitoring of manure storage facilities and put targets for cuts in nitrogen emissions into place in the 1990s. By comparison, efforts in the United States have been much less forceful. In 2009, the US Environmental Protection Agency ruled that CAFOs that emit 25,000 metric tons of greenhouse gases per year—which would affect about one hundred CAFOs (1 percent of the total) nationwide—would have to report their estimated emissions under the Greenhouse Gas Reporting Rule. But the United States then barred the EPA from mandating greenhouse gas emissions reporting or from requiring CAFOs to apply for Clean Air Act permits for greenhouse gas emissions. Since 2009, CAFOs have been able to skirt regulations under the

Comprehensive Environmental Response, Compensation and Liability Act (CERCLA), and only large CAFOs need to report emissions under the Emergency Planning and Community Right to Know Act (EPCRA). A 2012 Congressional Research Service report on animal agriculture and air quality estimated that livestock, including enteric fermentation and manure management, contributed 34 percent of methane and 4 percent of total nitrous oxide emissions from the United States. The American livestock industry, arguably to an even greater degree than those in other countries, continues to push back against efforts to limit livestock production's contributions to greenhouse gas emissions.[21]

Part of the issue with livestock production and greenhouse gas emissions is that so much of the problem rests on what many, especially in the livestock industry, see as natural processes. That is, enteric fermentation in domestic livestock, especially cattle, produces almost 70 percent of all the methane emanating from agricultural sources in the United States. To reduce methane production would mean reducing the number of cattle and other ruminants. Indeed, that is not only what commentators like Vaclav Smil have advocated, but what is also increasingly occurring around the globe as meat production shifts from beef to chicken. Chickens and turkeys are more efficient converters of feed to protein than cattle, and they produce much less greenhouse gas emissions. Some methane reductions from changes in cattle and sheep feed and rumen microorganisms are possible as well.[22]

Livestock manure contributes a part of the remaining portion of methane production. In several nations, particularly in China, India, and Western Europe, efforts have been underway for some time to capture the biogas from livestock manure to produce energy. Biogas released from the decomposition of organic matter normally consists of 40–70 percent methane with carbon dioxide making up most of the remaining gas. When captured through anaerobic digesters, the biogas can be used as fuel for heating, lighting, transport, and production of electricity. It can also be burned off, or flared, without useful human benefit. The first-ever anaerobic digester was built in 1897 in India and used to generate gas for lighting at the Matunga Leper Asylum in Bombay (Mumbai). The first manure-fueled biogas plant was built in Bombay in 1900. However, it was not until the 1950s that a variety of anaerobic digester plans were developed and disseminated. As of 2011, India has four million household biogas plants while China has 25 million household biogas plants, with about 10 percent of its rural households using captured biogas for a power source. In the West, as previously noted, anaerobic digestion was more typically used to treat wastewater and was not often used as source of biogas power. Covered lagoons capture biogas effectively but, in the case of the United States, have seldom been used to convert biogas into energy. However, in the past two decades, larger-scale digesters have been built in Western Europe, especially, and elsewhere to capture biogas for use as a power source. Today, the most efficient biogas production occurs when manure is combined with other wastes. Since the 1990s, increasing amounts of municipal solid waste in Europe is being converted into biogas. As of 2012, the world's largest biomass power plant running exclusively on (chicken) manure is in the Netherlands. It delivers enough electricity to power 90,000 houses. The Beijing Dequingyuan Chicken Farm treats 220 tons of manure and 170 tons of wastewater per day to generate biogas; and the Liaoning Huishan Cow Farm converts manure from 60,000 cows to produce biogas that provides power to 45,000 homes.[23]

While commentators like Simon Fairlie see a leveling off in the contribution from livestock production to greenhouse gases, much remains unknown about how the rapid rise in China's livestock and meat production affects and will continue to affect global environmental conditions. As explained in the following chapter, because new health concerns after the mid-twentieth century caused many affluent Western consumers to pursue meat avoidance strategies and therefore make discussion of environmental improvements more acceptable, the growing affluence of urban residents in China and many parts of the developing world made rapid strides toward matching Western meat consumption levels. The ability to purchase in urban settings meat like that which Westerners had enjoyed since the late nineteenth century rapidly became available to Chinese and other developing countries' consumers in the late twentieth century. To a great degree, the leveling off in Western meat production evident today is offset by the rapid increases in meat production among the large populations in China and the developing world. Improvements in environmental problems due to declining production and consumption of meat in the developed world may very well be increasingly nullified by greater meat production and consumption in the developing world.

# 9

## Your Safest Course Is to Let Meat Alone

In 1868, Seventh-Day Adventist Church co-founder Ellen White asserted that because meat increases the likelihood of cancer, tumors, and pulmonary diseases, and consequently weakens human physical, mental, and spiritual health, "your safest course is to let meat alone."[1] During the late twentieth and early twenty-first centuries, her sentiments about avoiding meat have found increasing support among many affluent Western consumers. While varying concerns about meat safety have been persistent throughout modern history in virtually all countries, it was not until the late twentieth century that worries about meat's potentially negative impact on human health gained backing from governments, physicians, and scientific authorities. From the late nineteenth through the early post–World War II era across the globe, authorities promoted meat as essential to human health. Indeed, it was often assumed that the more meat consumed the better human health would be. While meat eating continues to grow in popularity in the aggregate around the world today, especially in most of the developing world, consumers in the wealthiest nations have curtailed their meat consumption and have certainly modified their meat preferences.

Arguably even more significant among consumers who have limited their meat intake than fears about meat's environmental effects are concerns about meat's effects on physical well-being. Beginning especially during the 1970s throughout the Western world, governments and health authorities of various kinds began to publicize how meat contributed to the rising incidence of coronary heart disease and certain types of cancer. By the mid-1950s, nutritional deficiency diseases in the United States and much of the West had been largely eradicated. The so-called killer diseases of the heart and cancers replaced them as the leading causes of death. Increasing numbers of studies from the 1950s through the 1970s pointed at saturated fat and cholesterol, especially from meat and dairy products, as leading culprits in these diseases. These results were then promulgated in new nutritional standards starting in the late 1970s, though not without considerable protest on the part of the meat industry. Moreover, starting in the 1980s in Europe with the discovery of "mad cow" disease and then increasingly in the 1990s in many parts of the world with a dizzying array of other meat microbial outbreaks, consumers were repeatedly alerted to the dangers of meat consumption. Starting in the West, consumers turned especially to poultry in the hopes that it would provide a healthier meat choice. Consumption of beef peaked in the 1980s and has fallen worldwide since that decade. Nevertheless, overall global meat consumption has continued to climb, though it has started to level off and even decline in some Western countries. Since the 1970s, too, vegetarianism and veganism have gained increasing numbers of adherents. Indeed, not only those that are eaten, but animal products more broadly have also declined in importance. Around the world, perhaps most notably in East Asia, which only recently became a significant meat-eating part of the world, health advocates have urged a return to traditional diets that diminish the role of meat. This chapter highlights how public health findings about meat consumption have contributed to new concerns about its dietary role.

## Cholesterol, Saturated Fat, and Killer Diseases

Throughout the nineteenth and the first half of the twentieth centuries, meat consumption was generally encouraged and applauded as a significant contributor to human health. As explained in part II, meat eating had its critics, but the most prominent voices in the medical profession and governments around the globe championed meat consumption. Few British medical journals mentioned obesity until the 1930s. Links between coronary heart disease and obesity, for instance, were noted, but doctors' opinions varied widely about the nature of this relationship.

While Sir William Osler noted in 1908 that atherosclerosis affected the wealthy more than the poor, and Russian scientist Alexander I. Ignatowski demonstrated a connection between fatty diets and atherosclerosis in rabbits in a paper published in the same year, research on the effects of saturated fat and cholesterol on human health did not become prominent until the 1950s. Professor E. H. Ahrens, Jr., and his colleagues at the Rockefeller University in New York published a study on the relationship between vegetable and animal fats and cholesterol in 1954 and announced the link between saturated fats and cholesterol in 1957. Nutritionist Marion Nestle credits Ancel and Margaret Keys's book, *Eat Well and Stay Well*, first published in 1959, with alerting Americans to the dangers of coronary heart disease and its links to saturated fats and cholesterol.[2] The Keyses advised readers to "[r]estrict use of fats in dairy products and in ordinary meats."[3]

From the late 1950s through the late 1970s, many of the medical associations of the Western nations endorsed diets that reduced saturated fats and cholesterol. Starting in 1958, the United States National Health Education Committee endorsed such a diet, followed by the American Heart Association in 1961. Government medical boards in Finland, Sweden, and Norway all did so in 1968, followed by similar groups in New Zealand, the United Kingdom, Australia, West Germany, Canada, and the United Nations' FAO during the 1970s. While at least sixteen expert health and medical committees in the United States between 1968 and 1978 advocated for dietary reforms in saturated fats and cholesterol, the USDA continued to promote a nutritional guide that stressed the importance of eating quantities of foods from four basic groups—dairy, meat, fruits and vegetables, and breads and cereals—without regard for fats or calories or anything problematic about foods in general.[4]

A turning point in the US government's nutritional advice occurred in 1977 with the publication of the US Senate's Select Committee on Nutrition and Human Needs' *Dietary Goals for the United States*. Committee chair Senator George McGovern later attributed his decision to ask in 1968 for the creation of a special committee on nutrition to his watching a CBS report on the nation's hunger problems. While the committee addressed hunger and several other nutrition issues during its nine-year existence, McGovern later claimed that publication of the *Dietary Goals* was the committee's "furthest reaching and longest lasting" achievement.[5] Beginning in 1976, the committee responded to the "accumulation of scientific data linking diet and many of the Nation's major killer diseases."[6] During the March 1977 hearings that led up to the issuing of the *Dietary Goals*, committee member Edward Kennedy noted a poll by Dr. Kaare Norum published in the *Journal of the Norwegian Medical Association*, which found that 214 scientists from twenty-three countries advocated diets with fewer calories, less total fat, and less saturated fat to reduce the risk of heart disease. Kennedy also quoted Dr. Mark Hegsted from the Harvard School of Public Health that the "diet of the American people is becoming increasingly rich in meat and other sources of saturated fats and cholesterol. . . . [T]his diet which affluent people generally consume, everywhere, is associated with similar disease patterns, high rates of heart disease, certain forms of cancer, diabetes, obesity, . . . and are the major causes of disease and disability in the United States."[7] Several spokespeople for livestock, meat, and dairy interests claimed that the science linking meat to disease was still disputed. For instance, Wray Finney, president of the American National Cattlemen's Association said, "Respected scientists have shown that there is no valid basis for a claim that a reduction in beef consumption, including marbled beef, will result in less heart disease."[8] One such scientist, Dr. Robert Olsen from St. Louis University's School of Medicine noted, "It is ironic that beef, whose consumption in countries all over the world has been associated with the best overall health statistics, is now under attack."[9]

Two editions of the *Dietary Goals for the United States* were issued in 1977, partly because the first edition's blunt focus on reducing meat consumption raised so many concerns among livestock and meat producers. Although the second edition modified some phrasing about the types of fat to eat, it retained the same emphasis on reduction of fats and cholesterol. The second edition of the *Dietary Goals* urged a 25 percent reduction in total fat, from 40 percent to 30 percent of overall calories consumed. Among the many statistics included in the *Dietary Goals* were graphs showing that between 1909–13 and 1976, the average American's consumption of fat as a proportion of total calories had risen from about 32 to 42 percent. The *Dietary Goals* also recommended cholesterol consumption of no more than 300 milligrams per day. In terms of suggestions about food selection and preparation, the *Dietary Goals* advocated that Americans "[d]ecrease consumption of animal fat, and choose meats, poultry and fish which will reduce saturated fat intake." Further on, the *Dietary Goals* indicated that "[h]igh levels of fat, saturated fat and cholesterol most often enter our diets in the process of acquisition of animal protein." Resonating with the much earlier nutritional messages of Wilbur Atwater that linked food choices with financial frugality, the *Dietary Goals* noted that "[a] shift to the dietary goals outlined offers potential for significant reduction in food costs."[10]

While publication of the *Dietary Goals for the United States* resulted in affirmation from several government bodies, including Surgeon General Julius Richmond in 1979, the National Cancer Institute in 1979, and USDA and new US Department of Health and Human Services (DHHS) guidelines in 1980, its focus on reduction of saturated fat and cholesterol was not universally advocated in the United States.[11] In 1980, the Food and Nutrition Board of the National Research Council of the National Academy of Sciences issued its own report, titled *Toward Healthful Diets*, that said "there was no reason for the average healthy person to restrict dietary intake of fat or cholesterol."[12] The political fallout from the food industry's protests about the report resulted in the USDA's creation of the Human Nutrition Information Service. One result among government publications was a focus on phrasing in new guidelines that avoided mention of either "red" or "fatty" meats and an emphasis on "lean" meats. The FDA did not help matters initially either since it lagged in labeling fat content in foods until the early 1990s. Furthermore, suggestions from the *Dietary Goals* did not make it into a new USDA visual depiction of food goals until 1992 when the Food Guide Pyramid was promulgated. Since that time, more attention in revised USDA and DHHS guidelines has been focused on limiting high-fat processed meats, organ meats, and animal fats.[13] In October 2015, the International Agency for Research on Cancer, the cancer agency for the World Health Organization, released a report on the carcinogenicity of red and processed meat consumption indicating that red meat was "*probably carcinogenic to humans*," particularly in terms of colorectal cancer, and that processed meat was "*carcinogenic to humans*," in terms of causing colorectal cancer (italics in original).[14]

Americans' fat intake decreased somewhat after the late 1970s. Saturated fats decreased from 15 percent of calories in the late 1970s to 12 percent in 1990. By 1990, fat made up about 35 percent of the average Americans' calorie consumption. However, between 1990 and 1995, fat consumption rose slightly but remained at about 34 percent of total calories largely because Americans consumed more calories. In fact, the 1990s saw widespread reporting of an obesity crisis in the United States that could not simply be linked to fat consumption. While the new nutritional guidelines were probably one of the many factors influencing

American consumers' choices, other dietary shifts due to economic and cultural factors were underway. Between 1950 and 1970, butter consumption fell by half, plant oil consumption increased by 70 percent, egg consumption fell by 20 percent, and poultry consumption increased 200 percent. During the 1970s, butter consumption fell another 20 percent, plant oil consumption rose another 16 percent, egg consumption declined another 12 percent, and poultry consumption rose by another 25 percent. During the last part of the twentieth century, Western consumers, as well as many others around the globe, ate more meat but replaced beef with poultry and pork. As explained below, the advice to eat lean meat helped to continue to shift consumers' attention to poultry with nutritional results that have not noticeably improved health.[15]

## Meat Microbes

Consumer concerns about diseased meat have been long-standing. As described in chapter 4, concerns about diseased meat motivated state involvement in meat inspections, partly for the purpose of protecting consumers but also to facilitate the protection and extension of meat trade. Since the 1980s, however, most concerns about diseased meat have been about the microscopic, or microbial, invaders—generally bacteria or viruses—that pose serious health risks to those who consume them. Although microbes of various types were serious threats earlier, notably tuberculosis, parasites were often equally serious problems with diseased meat. As explained, the parasite *Trichinella spiralis*, which caused trichinosis, was perhaps the most important danger lurking in pork from the mid-nineteenth to mid-twentieth centuries and certainly caused considerable state intervention in the United States and Europe. Over the past thirty years, global efforts to contain, control, and destroy meat microbes have become one of the most significant consequences of the growth of meat production and consumption.[16]

The microbial event that initiated the modern era of meat disease concerns was the outbreak of bovine spongiform encephalopathy (BSE), or "mad cow" disease, which originated from a cow on Pitsham farm in Midhurst, West Sussex, England, in late 1984. The cow may have been infected as early as the winter of 1981–82. It was not until November 1986, however, that the cause of the cow's illness was determined to be BSE. Ultimately, scientists concluded that a prion infection caused BSE.

A prion is a protein that, unlike a bacterium or virus, does not have nucleic acids so is not considered a living organism. Prion infections cause transmissible spongiform encephalopathies, which is a family of brain diseases most commonly seen in animals as scrapie. First described in the eighteenth century in Europe but not understood as a transmissible disease until 1936, scrapie was widespread in sheep in Britain and much of Europe from the mid-1700s. It continued to increase in British sheep into the nineteenth century before declining toward the end of that century but did not disappear. In fact, in 1947 the USDA spent $3 million on eradicating the disease in sheep herds in Michigan infected by imported British sheep.[17]

The BSE crisis devastated the British agricultural economy. Between the late 1980s and 2004, 184,000 cattle infected with BSE were killed along with another five million that had consumed potentially infected animal feed, while export bans resulted in additional lost income. The total loss to the British beef industry was estimated to be $6.7 billion. The causes of BSE and the links between BSE and the human illness of Creutzfeldt-Jakob disease (vCJD) took several years to uncover. Within a year after identifying BSE, John Wilesmith, the epidemiologist who headed the Central Veterinary Laboratory near Surrey, where the first infected cow was examined, concluded that BSE was the result of feeding cattle contaminated meat and bone meal that included scrapie-infected sheep residue. Wilesmith's theory built on ideas taken from chance mutation theory that had developed in the 1970s. While the public was astonished that this practice occurred, feeding rendered animal products to both cattle and sheep had been common in Britain since World War II. Indeed, the British government required that a minimum of 5 percent animal-derived protein by weight be included in animal feed until 1953. Another factor in the presence of BSE was the emergence of lower-temperature rendering processes that had become commonplace before the 1980s. After a British committee was established in February 1988 to make recommendations about the BSE situation, a ban on ruminant protein in the making of animal feed was put into place on July 18, 1988. Nevertheless, the number of BSE cases in the United Kingdom did not peak until 1992 at 37,280. The United Kingdom banned specified bovine offals for human consumption and the use of ruminant protein in all animal feeds.[18]

Mad cow concerns triggered ongoing rounds of livestock and meat trade bans beginning in the late 1980s. The EU and the United States banned live cattle exports from the United Kingdom beginning in 1989, though some member EU nations and other countries outside the EU allowed exports from cattle herds certified BSE free. Nevertheless, EU restrictions did not target rendered animal products such as beef tallow or meat and bone meal. The United States and Canada did not ban such products until 1997, though Australia had done so in 1989 and New Zealand in 1996. Japan failed to do so until 2001. Between 1989 and 2003, twenty-three countries reported BSE infections in cattle. Although scientists from Britain and the United States had insisted that BSE could not jump to humans, the announcement of a new variant of Creutzfeldt-Jakob disease in ten young people in Britain in 1996 undermined this assertion. The similarity in clinical symptoms between livestock and humans and the easily recognizable lesions in the gray matter of the brain stem where it joins the spinal cord provided undeniable evidence of the new BSE variant.[19]

The BSE-vCJD crises set in motion major global trade and health developments. GATT in 1994, which had established the World Trade Organization, also established the Sanitary and Phytosanitary and Technical Barriers to Trade Agreements, which gave further validation to the Codex Alimentarius Standards created in 1963 by the FAO and World Health Organization to develop global standards for foodstuffs. The Codex Alimentarius provides international guidelines and recommended practices that are supposed to insure food safety and promote international food trade. These bodies became centrally important in Europe and elsewhere as nations developed new health standards during the unfolding of the BSE-vCJD crises. In July 1996, the European Parliament set up a committee of inquiry to examine the BSE-vCJD crises that resulted in a thorough overhaul of the older system of food safety in the EU, which had allowed a decentralized response. Earlier alarms and resultant trade barriers in Europe over the use of hormones in US imported beef in the 1980s had started the process of centralization, but implementation of the European Parliament's Regulation no. 178 in 2002 completed this process, including establishment of the European Food Safety Authority. Since 2002, the EU has used Hazard Analysis and Critical Control Point (HACCP) practices to curtail food illnesses. Developed initially in the US private sector in the 1970s, HACCP focuses on seven control points: identifying and analyzing hazards; determining points in the process where hazards occur; establishing tolerance levels; establishing monitoring procedures; making corrections if deviations occur; keeping records; and verifying the HACCP plan. EU's food safety regime relies on microbiological testing and has implemented traceability features to try to ensure food safety. Although initial projections pegged the number of potential vCJD cases at 2,600 in 2003, the total number of deaths to date has been fewer than one-tenth that number, with more deaths increasingly unlikely given the eight-year incubation period between BSE and vCJD and the near disappearance of new BSE infections since 2004.[20]

Japan's reaction to the BSE-vCJD crises was slower than the EU's but was also motivated by similar trade and public health concerns. As explained in chapter 6, Japan's beef trade with the United States was severely curtailed from 2004 to 2006 before it started to recover in 2007. It was not until 2010 that Japanese imports of US beef started to pick up again. The US reaction to the BSE-vCJD crises was mild by comparison to the EU's and Japan's. After the discovery of BSE in a cow in the state of Washington in December 2003, authorities emphasized that the animal had been born in Alberta, Canada. In January 2004, the USDA announced it would test BSE in just 40,000 cows out of 35 million slaughtered each year. The random sampling policy was announced even in the face of international bans on beef exports from the United States from forty countries. After years of resistance and delays, the USDA approved use of HIMP—a program of HACCP for meat and poultry industries—in a few slaughterhouses in 1997 but did not require companies to test for pathogens until the early 2000s. Compared to the EU's emphasis on microbiological testing, the USDA still relied more extensively on visual inspections and practices that dated back to the Meat Inspection Act of 1906. Indeed, the United States did not more fully institute HIMP, as noted below, until *E.coli* O157:H7 and *Listeria* concerns became greater in the early 2000s. Food traceability, a key component of the EU's comprehensive food safety reorganization, was implemented largely in the United States as part of the response to the 9/11 crisis through the US Bioterrorism Act. The United States tended to see BSE as an economic hurdle rather than as a significant health threat, partly because through 2010 only two cases of BSE had been detected through the country's random sampling policies.[21]

Compared to its tentative response to the BSE-vCJD crisis, US public health authorities responded more vigorously to new threats from a variety of food-borne bacteria. The starting point for much of this new concern and reaction was the severe outbreak of *Escherichia (E.) coli*

O157:H7 from Jack in the Box restaurant hamburgers that started in Los Angeles in November 1992 and spread throughout the West Coast by January 1993. During the span of this outbreak, 732 people became ill, 195 were hospitalized, 55 developed hemolytic uremic syndrome, and four died from hemolytic uremic syndrome. The public relations fallout from the crisis and the resulting lawsuits nearly bankrupted Jack in the Box. *E. coli* is a type of bacteria that naturally inhabits the intestines of many mammals including humans, but the O157:H7 variety that is common in livestock is deadly in humans. The bacteria are found in cattle fecal matter and kill humans by damaging cells that line the intestine. The O157:H7 variety is believed to have evolved from another *E.coli* variety, O55:H7. The first outbreak of illnesses occurred in the United States in 1982 from contaminated hamburgers. Between 1982 and 1993, a total of twenty-four Americans and Canadians died from O157:H7 infections. Although a problem not confined to meat, since cross-contamination with fecal matter can occur with many food products, between 1994 and 2005 the USDA issued 145 recalls of over 64 million pounds of contaminated hamburger. After the Jack in the Box outbreak, the USDA required all restaurants to cook hamburgers to an internal temperature of 155, later 160, degrees Fahrenheit. The USDA's HIMP procedures included pathogen testing in the early 2000s. Even so, the American meat industry and the USDA continue to emphasize that pathogen problems in meat are best addressed by individual consumers who use safe handling and proper cooking procedures.[22]

Although *E.coli* O157:H7 infections have generally received the most attention, other bacteria are actually more common sources of foodborne illnesses in the United States In 1999, the two most common sources of bacterial illnesses in the United States were caused by *Campylobacter jejuni* and *Salmonella*. Both are commonly found in the intestinal tracts of livestock, especially poultry. A 2003–2004 survey carried out in New Zealand found that 89 percent of chickens tested positive for *Campylobacter*. As part of that same survey, 68 percent of chickens in Taiwan, 59–75 percent in the United Kingdom, 50 percent in Germany, 46 percent in Japan, 25–40 percent in Denmark, and 36 percent in Mexico also tested positive. *Campylobacter* is highly sensitive to heat and cold and is killed when heated or frozen. Although *Salmonella* is also killed when exposed to heat, salmonellosis is nevertheless common not only in the United States but in many other parts of the world as well. In 2000, per 100,000 people, Denmark reported 43 incidences,

Australia reported 32 incidences, and England and Wales reported 25 cases. Although much less common than either *Campylobacter* or *Salmonella*, other bacteria that cause severe illnesses include *Clostridium perfringens*, *Listeria monocytogenes*, *Shigella*, *Yersinia enterocolitica*, and the *Vibrio* species. All except the *Vibrio* species, which are found in fish and shellfish, are commonly found in livestock intestines and feces. *Clostridium botulinum* is rare but especially deadly and is the source of particular concern as a possible bioweapon. Also alarming are the increasing number of infections due to antibiotic resistance traced to use of animals raised in confinement, perhaps most notably MRSA, or methicillin-resistant strain of *Staphylococcus aureus*. Deaths in the United States due to antibiotic-resistant bacteria doubled between 1999 and 2005 from 11,200 to 23,000. Although less publicized, food-borne illnesses caused by protozoa, which are single-cell eukaryote animal-like creatures, such as *Giardia lamblia* and *Toxoplasma gondii*, are fairly common worldwide. Moreover, the most common food-borne illness—estimated to be 58 percent of all such illnesses in the US—is actually caused by a virus, norovirus, which, because it causes mild and brief symptoms, is typically not reported.[23]

## Modern Meat Avoidance Strategies

With an estimated 76 million cases of illness, 323,000 hospitalizations, and 5,200 deaths attributed to food-related causes each year just in the United States, it is no wonder that consumers around the globe have pursued a variety of meat avoidance strategies. Long-term health concerns about fat and cholesterol, meat microbes, as well as potential threats from agricultural chemicals in pesticides, herbicides, and fungicides and the rapid proliferation of genetically-modified foods, have made consumers around the globe, especially in the developed world, much more wary of meat eating. Added to this is the growing concern about animal ethics. Vegetarianism reemerged in the West starting in the 1960s, and veganism has also garnered larger numbers of adherents. More important than either movement has been the growing number of semivegetarians, or flexitarians, a term first used in 2003. The use of animal products other than meat has also dropped during the last fifty years, partly for reasons having to do with meat avoidance strategies.[24]

Western vegetarianism of the past half-century exhibits many similarities with the movement that emerged in England during the early nineteenth century. Contemporary vegetarianism in the West shares the threads of health and hygienic concerns as well as the ethical and humanitarian emphases that infused the nineteenth-century movement. Estimates of the number of vegetarians in North America range between 1 to 3 percent, with a slightly higher number in Europe. Some surveys have found that as many as 7 percent of North Americans identify as vegetarian, but typically many of these self-identifiers also either eat meat occasionally or do not count fish or poultry as "meat." Indeed, ovo-lacto vegetarian status—no consumption of flesh but the eating of eggs and dairy products is acceptable—is defined as the baseline definition for vegetarianism while pollo- or pesco-vegetarianism, consumption of poultry or fish but no red meat, is defined as a type of semivegetarianism. Following the BSE and vCJD scare, the number of vegetarians in the United Kingdom in 2001 was estimated to be as high as 12 percent. While certainly a relatively small percentage, the number of vegetarians in North America and Europe today is much higher than it was during the nineteenth century. Indeed, in terms of its broader cultural acceptance in Western foodways, vegetarianism is mainstream in a way that was never true in the nineteenth century.[25]

Vegetarianism outside North America and Europe is primarily found in South Asia. In India, about one-third of the population is vegetarian, particularly in the south and in Gujarat in the northwest. Unlike the West, where nonconformist, perfectionist Christian traditions associated with vegetarianism such as the Bible Christians, Quakers, Unitarians, and Seventh-Day Adventists currently have very little to do with contemporary meat avoidance advocacy, Hindu, Buddhist, and Jainist traditions are still very influential in India's vegetarianism. All three traditions share a focus on *ahiṃsā* (nonviolence). Of these traditions, Jainism is certainly the most thoroughly vegetarian in its teachings. Yet the approximately 4 million Jains in India comprise only 0.4 percent of the country's population. Buddhist traditions advocate that animals not be hurt but do not insist upon vegetarianism. Indeed, throughout Buddhist history, ethical interpretations allowed people to consume meat as long as the meat was not directly killed for the consumer. Among Hindus, the Brahman caste has tended to have the most vegetarians but often more as a way to avoid caste pollution and less about health or animal ethics. In contemporary India and Sri Lanka, cow veneration and vegetarianism are also part of anti-Islam *halāl* abolitionist movements and expressions of Indian nationalism. In China, vegetarianism is primarily a remnant of rural poverty, while vegetarianism has largely disappeared in Japan.[26]

Although vegetarian advocates and researchers disagree about the primary motivations of those who become nonmeat consumers, both health and ethical concerns are important motivations. In North America, most vegetarians are educated, middle-class women. Some surveys suggest that 70 percent of vegetarians are women. Many young people also claim to be vegetarian as well. Vegetarians in the United States received support for their nonmeat diets beginning in 1980, when the American Dietetic Association began to issue periodic position papers on vegetarian diets. Beginning in 1995, the US government's dietary recommendations included vegetarianism as an acceptable option. During the 1990s, the ADA and other government reports also debunked the need for so-called complementary proteins—first articulated in Frances Moore Lappé's *Diet for a Small Planet*, published in 1971—and dismissed the need for dairy products in healthy diets. Thus, in the vegetarian food pyramid, nutritional needs are met through a combination of whole grains, legumes, and soy foods, combined with vegetables and fruits and small amounts of nuts and seeds and vegetable oils.[27]

The most ardent antimeat and anti–animal product segments of North American and European consumers are vegans. The Vegan Society was founded in England in 1944 by Donald Watson as a splinter group of vegetarians. Watson's newsletter was circulated in the United States before H. Jay Dinshah founded the American Vegan Society in 1960 to promote Jainist beliefs about *ahimṃsā*. Although estimates vary widely, about a third as many vegans as vegetarians can be found in North America and Europe. While vegans share vegetarians' motivations in terms of health and ethical concerns, vegans generally express stronger beliefs about animal ethics. Animal liberation is often the goal of vegans and underscores their concerns about the need for a more thorough transformation of not only human diet and health standards but market capitalism, agricultural practices, and ecological standards more generally. Vegans in North American and Europe carry the mantle of utopian reform in a more thorough-going way than virtually any nineteenth-century vegetarian reformer could have imagined.[28]

Although it is difficult to directly connect vegans' rejection of any use of a product containing animal parts, the use of animal by-product items has declined in North America. Most dramatic has been the declining demand for edible by-products since World War II. While the US meat industry exports huge amounts of animal hides, fats, and by-product meats, by the end of the twentieth century few meatpackers saved all the edible offal found in cattle, hogs, and sheep. Use of animal-derived pharmaceuticals peaked in the late twentieth century and has been on the decline. For instance, insulin from cattle pancreases has been increasingly replaced by biosynthetics such as Humulin. Since the BSE scare, the United States and other countries around the world no longer use mechanical deboning techniques to remove meat from around cattle spinal cords.[29]

While vegans are centrally concerned with ending speciesism, a term first popularized by Peter Singer in *Animal Liberation*, it is not clear how concerned many animal rights groups are with promoting vegetarianism or veganism. The world's largest animal rights organization, People for the Ethical Treatment of Animals (PETA), founded in 1980 by Ingrid Newkirk and Alex Pacheco in the United States, is a main target of those who find fault in its health and animal rights approaches. In particular, many vegans accuse PETA and other animal welfare organizations of pursuing approaches that—while meant to make consumers more comfortable with notions of humanely raised livestock—allow continued exploitation of animals for profit.[30]

Despite the fact that they receive support from most of the Western medical and government health establishment, both vegetarians and vegan dietary regimes have come under attack by groups espousing diets that reject low-fat emphases. Supporters of the work of the Land Institute's promotion of perennial crop polycultures, such as Michael Pollan and Lierre Keith, emphasize how a diet based on grains like rice, wheat, and corn both degrade the natural environment and cause severe health problems. Other critics of grain-centered diets include those who emphasize the composition of ancient humans' diets. The so-called Paleolithic, or Paleo, diet claims that low-fat meat, fish, root vegetables, and fruit should form the basis of a healthy diet. These diets, which emphasize that carbohydrates not fats cause heart disease, diabetes, and most cancers, also reject the dietary standards that most Western governments have espoused for more than a generation.[31]

Undoubtedly, the most important Western meat avoidance strategy has been semivegetarianism, especially in terms of the substitution of poultry for beef. As beef consumption stagnated or fell, poultry consumption has grown by more than 500 percent over the past fifty years. Early on in the post–World War II era, nutrition experts touted poultry as healthier than beef.[32] Ancel and Margaret Keys's *Eat Well and Stay Well*, noted explicitly that compared to beef and pork, "[c]hicken is very different. Chicken meat contains much less fat, especially if the birds are young."[33] The *Dietary Goals for the United States* endorsed chicken as well: "Decrease consumption of animal fat, and choose meats, poultry and fish which will reduce saturated fat intake."[34] Those with ethical concerns about meat eating responded to rising alarm over BSE in cattle (which coincided with the decreasing cost and improved marketing of poultry) by deeming it more acceptable to eat birds than mammals.[35]

## State Promotion of Traditional Food Cultures

In much of North America and Europe, governments have promulgated dietary advice cautioning against excessive meat consumption. In Japan and China, authorities have encouraged their citizens to return to many elements of their pre–World War II diets as a means of promoting better health in the face of rising rates of disease associated with growing affluence. By the early 2000s, Japanese health authorities estimated that about one-quarter of its population was obese. About one-third of men aged 40 to 49 were found to be obese according to the Japan Society for the Study of Obesity. In June 2005, Japan enacted the Basic Law on *Shokuiku* (food education). According to the Ministry of Agriculture, Forestry, and Fisheries' information on the law, it was passed because of growing concerns about obesity and other lifestyle-related diseases. The Ministry claimed that, around 1980, a fundamental shift in Japanese diets toward more fat than in traditional diets resulted, for instance, in 30 percent of men between the ages of 30 and 60 being overweight.[36]

The transition from grains and vegetables that predominated in Japan's diet in 1900 to the much more Western array of protein, fats, grains, fruits, and vegetables one century later was nothing short of remarkable. At the end of the nineteenth century, Japan's dietary composition was 60 percent rice, 18 percent potatoes, 16 percent barley, 5 percent cereals, and about 3 percent foods of animal origin. By 1950, animal protein and fat still composed no more than 5 percent of the average Japanese diet. By 1963, though, the percentage of animal foods made up about 10 percent of the Japanese diet; this figure rose to 20 percent by 1977 and has been at about 25 percent since the mid-1990s. This compositional shift occurred without any increase in total overall caloric input. Vaclav Smil and Kobayashi Kazuhiko emphasized just how remarkable this dietary transition was: "the average Japanese diet [has] seen more profound relative changes during the fifty years after 1950 or 1955 than the American diet did during nearly the century between 1909 and 2006."[37] Researchers have also shown that Japanese and other East Asian countries' peoples grew significantly in height over the course of the twentieth century largely attributed to the increasing availability of animal protein in their diets.[38]

While consumption of increasing amounts of animal protein and fats resulted in improvements in Japanese health and stature, this dietary shift also created many of the same disease risks that plagued Western cultures during the late twentieth and early twenty-first centuries. The seven aims of the Basic Law on *Shokuiku* illustrate not only health concerns but reminders about Japan's longstanding support for economic self-sufficiency and protection for domestic farming and fishing:

1. *Shokuiku* should have the purpose of promoting people's health in body and in mind, as well as enriching human lives;
2. Care should be taken to induce people to develop greater appreciation for and understanding of their diets, including the various roles played by the natural environment and the people who produce, transport and prepare food;
3. Nationwide, voluntary movements for promotion of *Shokuiku* should be developed;
4. Especially [p]arents, educators, and daycare providers should actively promote *Shokuiku* among children;
5. Understanding of dietary issues should be reinforced by taking advantage of all available opportunities, such as at home, in schools, and in the community or everywhere, to offer a variety of food-related experiences and activities;
6. Awareness and appreciation of traditional Japanese food culture as well as food supply/demand situations should be promoted, and opportunities of interaction between food producers and consumers should be created, in order to revitalize rural farming and fishing regions, and to boost self-sufficiency in Japan;
7. Appropriate information on food safety should be offered, in order to help people to practice proper diets.[39]

The Japanese Food Guide Spinning Top that accompanied the *Shokuiku* efforts arranged food groups inversely from the US food pyramid (which was replaced in 2011 with a MyPlate diagram) so that the highly recommended category of grains and vegetables was placed on top; fish and meats in the middle; and milk, dairy products, and fruit on the bottom. Snacks, sweets, and beverages other than water or tea were placed off to the side. Compared to the US food pyramid of the same period, the Japanese Food Guide Spinning Top recommended that consumers eat more vegetables than Americans and somewhat lesser amounts of fruits. Japanese consumers could eat relatively more fish and meat and less milk and dairy products than their American counterparts. In the commentary portion of the Spinning Top, Japanese consumers were advised to "[b]e careful not to eat too much fish and meat dishes. Especially, greasy dishes could result in excessive fat and energy intake."[40] The Japanese government projected that, by consuming this recommended mix of food, food sufficiency levels—which on a production value basis had fallen from an overall 86 percent in 1965 to 70 percent in 2005—would return to 76 percent by 2015.[41]

Chinese health authorities have also started to take note of their people's rapidly changing diets and the negative effects on public health. Ironically, one of the largest epidemiological studies in history focused on the good health of rural Chinese and the lessons that could be learned by Westerners from the traditional Chinese diet. The China Study, or China-Cornell-Oxford Project, conducted by the Chinese Academy of Preventive Medicine, Cornell University, and Oxford University over a period of about twenty years, began in 1983. The study looked at rates of cancer and other chronic diseases from 1973–75 data in 65 counties in China and correlated those data with 1983–84 dietary surveys and blood work from one hundred people, ages 35–64, in each of the counties. The assumption was that because these populations and diets had changed little over generations, current health histories and past death rates could be viewed as evidence of the impacts of long-established dietary differences. The researchers, including Dr. T. Colin Campbell from Cornell University, wrote a very popular health book, *The China Study*, first published in 2005. It concluded that populations of counties with high consumption of animal-based foods in 1983–84 were more likely to have had death rates from cancer and chronic diseases during the period 1973–75 that were higher than death rates in counties in which the diets comprised more plant foods. Campbell claimed in *The*

*China Study* that the research indicated that "relatively small intakes of animal-based food were associated with adverse effects."[42] Based on the China-Cornell-Oxford Project findings, *The China Study* recommended a diet consisting of fruits, all kinds of vegetables (including legumes and nuts), and whole grains. The book advised eating only minimal amounts of refined carbohydrates like pasta, vegetable oils, and fish, while urging complete avoidance of red meat, poultry, dairy, and eggs. In other words, *The China Study* urged that people pursue a near vegan diet.[43]

While *The China Study* lent support to vegetarian and vegan dietary positions in the West, the Chinese people—beginning especially in the 1990s—largely abandoned their traditional vegetarian diet. Especially as China's urban population exploded in the late twentieth- and early twenty-first centuries, China became "Fat China." As Paul French and Matthew Crabbe note in their book with this title, the Chinese adopted Western-style diets as their incomes grew. The 2015 Report on Chinese Nutrition and Chronic Disease, released by China's National Health and Family Planning Commission, found that hypertension, diabetes, cancer, and heart disease also increased significantly between 2002 and 2012. By 2012, 86.6 percent of all deaths were caused by chronic diseases, with heart disease, cancer, and lung disease the leading causes of death. The commission noted that smoking, overdrinking, lack of physical exercise, and poor diets that included too much salt and fat were the causes of these chronic diseases. It seems likely that China, like Japan, will soon urge its citizens to return to elements of its own, largely nonmeat, traditional diet as a means of restoring public health.[44]

If Chinese authorities join Japan's in supporting diets less focused on meat, then East Asian meat consumption could start to level off sometime during the twenty-first century, as Chinese and Japanese populations join Western consumers, who have cut back on the amount of beef, pork, and mutton in their diets in the half-century since 1961. Meat consumption in much of the West would actually be roughly the same or even less than it was in 1961 if not for the explosion in poultry consumption. The erosion in meat eating other than poultry suggests that the efforts of Western physicians, scientists, and government authorities in voicing concerns about the effects of meat consumption on human health has made a difference in Westerners' views of meat. Arguably just as important has been the growing concern about meat microbes. And while the number of vegetarians and vegans in the West is still relatively small, the diffusion of their ideas about both health and animal ethics seem to have contributed to the growth of semivegetarianism. Much of the future of meat eating around the globe in the twenty-first century may depend on the ways in which Western concerns about meat eating influence the growing number of urban, affluent consumers in the rest of the world.

# 10

## Meat in the Twenty-First Century

During the latter half of the nineteenth century, meat, a significant dietary component of only the wealthy in the West for hundreds of years, fast became a much more prominent part of the diets of most Westerners. As Western economic, political, and cultural influence expanded around the globe during the late nineteenth and early twentieth centuries, beliefs about the importance of meat not only as a signifier of status but also as essential for human health gained adherents outside the West. Science—including pseudoscientific social theories such as eugenics—and state policies and promotions helped to legitimate the increasing consumer demand for meat in the West and to convince non-Westerners that meat was essential to their health and prosperity. Protein's discovery, scientific assertions that animal protein was vital for good health, and the marketing of beef in the form of inexpensive extracts and canned products created consumer demand among all social classes. State intervention in controlling livestock diseases and, especially, zoonotic diseases transmitted from livestock and meat to human beings reassured people that meat, often feared as dangerous, was safe to eat. Other state regulations of meat production created increasing amounts of meat for both domestic consumption and foreign trade among Westerners through the first half of the twentieth century. As Western meat production and consumption increased, consumers expressed a growing desire for beef, the meat that had traditionally held the most prestige and status value. The meat industry also created new demand for animal products through systematic marketing of inedible by-products. By the mid-twentieth century, Westerners considered meat and animal by-products to be invaluable, and many outside the West also embraced these developments.

After World War II, non-Westerners' demand for meat was increasingly realized. The second food regime facilitated meat's global expansion in two major ways. First, North America provided the feed grain necessary for European and Asian nations to build up their own livestock and meat industries. Through the 1990s, the United States dominated this feed grain trade. Second, the West provided meat to the growing numbers of affluent consumers in non-Western nations, which further reinforced the global expansion of meat consumption. Yet even as nations such as Japan and China joined the world's important meat cultures by the end of the twentieth century, a variety of developments, most notable in the West, created growing concerns among affluent consumers that meat may not be as essential to human health as believed.

Westerners' concerns focused on three major areas: workplace discrimination and, especially, environment and human health. After a long history of prejudicial treatment of workers involved in meat production due to stigmas about their involvement in a killing industry, state intervention over the middle third of the twentieth century created more protections and benefits for them. But state protections for meat workers eroded worldwide as the second food regime faded and transitioned into the current corporate-dominated food regime. Increasingly, even as many recognized the plight of meat and other food workers, consumers became even more alarmed by evidence of meat workers as violent and abusive, particularly in light of the animal rights movement that has developed since the 1970s. Concerns about the effect of livestock production on the environment and meat consumption on human health have helped to curtail beef consumption in the West, though poultry, a kind of meat seldom eaten until the late twentieth century, has continued to gain popularity. Poultry production and consumption are less threatening to the environment, since the industry contributes much less greenhouse gas to the atmosphere than cattle production; and poultry is also touted as healthier than beef, though that is often not true depending on its preparation. Although meat production has increased steadily on a global scale, it has already leveled off to a great degree in much of the West, especially for meats other than poultry.

Important indicators show that many Westerners are choosing dietary paths as semivegetarians, vegetarians, or vegans, and governments are increasingly advocating limits on meat consumption or recommending pursuit of traditional, more limited meat diets.

While consensus seems to be growing among affluent Westerners that meat eating should be curtailed, many observers nevertheless see a near future of continued expansion of meat production and consumption. Thoughtful commentators about meat, such as Vaclav Smil, note that "there does not appear to be anything in the foreseeable future that could fundamentally change today's practices of growing livestock for meat."[1] Smil explains that while meat eating has seemingly reached a peak in much of the developed world, demand is rising among people in Asia, Africa, and Latin America, which will likely offset the stagnation or declines in Europe and North America. He notes that vegetarianism and veganism still constitute only very small percentages of the affluent populations in the West and that sales of meat substitutes and development of *in vitro* meat are not likely to make much of a dent in overall meat production and consumption.[2]

If Smil and others are right, then humanity faces the grim prospect of coping with human and animal health and environmental issues that are troubling to say the least. The FAO projects that overall meat consumption will increase by 43 percent between the recent period 2005–2007 and 2050, with about 70 percent of all meat production by mid-century in currently developing countries. In order to meet that level of demand, biologist Colin Tudge estimates that the number of livestock animals needed for slaughter worldwide each year might reach 120 billion—double the current 60 billion—all of which would consume the equivalent of 4 billion people's food. Effectively, then, the Earth's population in 2050 would be 13.5 billion instead of the projected 9.5 billion. The effect of this near doubling in meat production and consumption on an increasingly fragile planet is frightening to say the least.[3] Despite his assertion that livestock production will continue, Smil concludes his study of the role of meat eating in human societies on a hopeful note. "There is a high probability that by the middle of the 21st century global meat production will cease to pose a steadily growing threat to the biosphere's integrity."[4]

Smil's argument is predicated on the notion that humanity will pursue rational and efficient meat production and consumption. But what are the possible avenues through which this rational and efficient level of production and consumption might occur? One avenue is through creation of meat substitutes, including in vitro meat, fake meat, and entomophagy. *In vitro*, or cultured, meat has been created in laboratory settings using cells cultured from animal tissue. The first cultured hamburger was cooked and eaten at a press conference in London in 2013. Though cultured meat is currently enormously expensive to produce, some researchers estimate that the cost may be comparable to real meat in another ten years. Fake meat, made with plant proteins such as soy, has also increased in popularity, though the taste is often thought to be lacking and the cost uneconomical. Another long-standing meat substitute in many parts of the world is entomophagy, the eating of insects. Likely an essential source of protein for early humans, insects were also very significant and continue to be important, at least seasonally, in many indigenous cultures around the world, particularly in Southeast Asia, South Asia, Papua New Guinea, central and south Africa, and Mexico. Today in Kinshasa, Democratic Republic of Congo, 96 tons of caterpillars are consumed annually, about 11 ounces per week per capita. At least 1,900 species of insects on the planet are known to be edible. Although insect production has tremendous environmental benefits over livestock production and many species of insects have the equivalent or even better overall nutritional value than red meat, its proponents admit that Western peoples are generally appalled and, at best, mildly curious about insect consumption.[5] Daniella Martin, one of the popular advocates for entomophagy, encourages Americans to try insects since they "tend to taste a bit nutty, especially when roasted."[6] Scientists such as Arnold van Huis see promise in extracting insect proteins for human food products. He and others see as even more promising the efforts underway in Europe and North America to develop edible insects for poultry and fish feed, partly because those animals naturally consume insects. In addition to Westerners' squeamishness, another barrier is the lack of guidance in most countries for use of insects as human food. For example, the Codex Alimentarius does not include insects in its regulatory framework.[7]

One way that consumers' behavior about eating meat might be modified is through a meat tax. Since 2014, increasing numbers of environmentalists and animal ethics proponents have called on governments to impose a carbon tax on beef production since it clearly contributes to greenhouse gas emissions. The idea has some support among economists as well, though many others see such a tax as a burden to lower-income consumers and farmers. In early 2016, the Danish Council of Ethics approved the concept of taxing beef production, and the measure was sent to the Danish government for consideration. Given the antitax mindset of many Western voters, it does not seem likely that new taxes on meat will find much traction in the near future. But given that there seems to be a growing groundswell of concern about meat production's effects on people and animals, greater educational emphasis on developing support for agricultural sustainability may prove to be a more long-lasting way to gain broader acceptance for any of the possible meat substitutes.[8]

The concept of sustainability originated as both a developmental perspective and policy goal in the United Nation's World Commission on Environment and Development (or Brundtland Commission) Report, *Our Common Future*, issued in October 1987. The report used the concept of sustainable development to focus on a wide array of global problems, including agriculture. Sustainability—in regard to agriculture as well as a broad array of other complicated issues—uses a systems approach that emphasizes the connections and interactions among a broad set of social, economic, and environmental factors.[9] Indeed, even van Huis's and Martin's arguments about the value of entomophagy are based largely on the environmental sustainability of insect production and consumption over livestock. Joan Dye Gussow has noted that the underlying problem of modern meat production is "a problem of scale and control" and that what is needed is combining concerns about human and animal rights with a sustainable agriculture movement.[10] The most practical environmental and public health response to increasing global meat production and consumption is likely not complete eradication of meat production and consumption but building consensus for a more reasonable balance.[11]

Education about sustainable livestock production could start with reminders about the value of all lives—people and animal—and, ultimately, how human and animal lives are connected in a cycle of life and death and go on to increase awareness of the deep traditions in Eastern and Western history that promote livestock production based on such considerations. Ancient Buddhist traditions throughout Asia promoted ethical considerations of animal, especially cattle, killing. Two hundred years ago in Britain, James Plumptre's *The Experienced Butcher* not only explained the need to kill livestock mercifully but reminded readers that "humane and responsible butchers do exist." He emphasized that the problem with meat is fundamentally a matter of human tastes: "in general . . . those who have the means of eating animal food, commonly eat too much."[12] David Perkins maintains that Plumptre, like Romantic authors throughout the nineteenth century, connected treatment of animals to "a much broader civilizing, educative, and disciplinary effort."[13] In a manner that harkened back to this Romantic tradition, British animal welfare advocate Ruth Harrison's *Animal Machines*, published in 1964, condemned intensive farming methods of broiler chickens and veal calves: "A danger of accepting any form of life as cheap is that each successive generation might accept slightly lower standards."[14] Peter Singer's *Animal Liberation* attributed much of his motivation for writing the book—and many of his examples in the book about livestock abuse—to Harrison's work. Singer advocated vegetarianism or veganism as the best path to alleviating livestock animal suffering. Kate Rawles, a contemporary commentator on sustainability, advocates the addition of animal welfare to the systems approach about environment, society, and economy that underpin sustainability models. Specifically, she urges adoption of the UK's Farm Animal Welfare Committee's Five Freedoms—Freedoms from Hunger and Thirst; Discomfort; Pain, Injury, or Disease; Fear and Distress; and Freedom to Express Normal Behavior—as the principles of animal welfare.[15]

Compared to intensive livestock practices, which stress efficiencies and low monetary cost, sustainable livestock production emphasizes that nothing is cheap about the making of meat. Sustainable livestock production is also predicated on three main historical benefits of animal farming: provision of food and other animal products that contributed positively to people's lives and livelihoods; livestock's role in both natural and human-managed nutrient and energy cycles; and ruminant livestock's ability to use biomass on lands otherwise not suitable for plants that humans can consume. Regarding livestock's role in nutrient and energy cycles, Tony Milligan has pointed out that universal vegetarianism and veganism both confront humanity with quandaries. Universal vegetarianism ignores the fact that vegetables and fruits are not grown readily in many places on the planet. In addition to the issue of vegetable and fruit production, universal veganism would require a return either to widespread use of human manure as fertilizer or even greater, and heavily polluting, use of chemical fertilizers. Nicolette Hahn Niman has expounded on cattle's important contribution to productive use of marginal pasture lands.[16]

Building on the characteristics of sustainable livestock production, epidemiologists Anthony J. McMichael and Ainslie J. Butler highlight three additional steps: reduction in total global consumption of meat; changes in livestock husbandry methods; and changes in the overall profile of preferred livestock species and strains. Clearly, the first point is the most challenging to achieve. Among the many elements associated with overall global meat consumption today are the connections among education, affluence, and declining interest in meat in contrast to the situation in the nineteenth and early twentieth centuries in the West. Observers in both China and Japan, where, especially in the former, meat consumption continues to increase, see some development of consumer awareness of the environmental and health consequences of increased meat consumption. Japanese researchers have placed greater emphasis on poultry and pork instead of beef for more environmentally efficient meat production. Colin Tudge advocates application of "enlightened agriculture," in which people would eat plenty of plants, not much meat, and would pursue maximum dietary variety—all elements, as he notes, that are traditionally present in what he claims are the world's greatest cuisines of Provence, Italy, Turkey, Persia, India, and China. In each, meat was basically a flavor enhancement instead of a staple. The increasing numbers of educated, relatively affluent semi-vegetarians suggest a possible path toward achieving meat consumption reductions.[17]

Regarding livestock husbandry methods and livestock species and strains, considerable research is now being conducted on genetic, nutritional, and animal health improvements, as well as new emphases on mixed animal-plant farming systems that had characterized farms in the West before World War II. The report of the US National Research Council of the National Academies' Committee on Twenty-first Century Systems Agriculture, *Toward Sustainable Agricultural Systems in the 21st Century*, highlights the need for US agriculture to achieve sustainability through both incremental and transformative approaches. Building on its 1989 report, *Alternative Agriculture*, which was one of the earlier statements about the use of a systems approach in farming, the 2010 report stressed that scientific, public policy, and marketing developments and decisions would all be crucial in developing these sustainability approaches. Regarding livestock farming, the report outlines both alternative and mixed crop–livestock approaches. For instance, alternative systems would rely more on pasturing and less on confinement of livestock, while mixed crop–livestock farming would require livestock farmers to produce more of their own animal feed requirements.[18]

Is sustainable livestock and meat production possible? In 1894, Henry Salt raised the question regarding animal rights that is worth stating here in regard to sustainable livestock production: "But how is this vast educational change to be inaugurated—let alone accomplished? Like all far-reaching reforms which are promoted by a few believers in the face of public indifferentism, it can only be carried through by the energy and resolution of its supporters."[19] Supporters of sustainable livestock and meat production certainly exist. College and university programs in sustainability are increasing around the world and are training a whole new generation of leaders who will apply these ideas to agriculture and other complicated systems. Meat has made people powerful, but now all people need to consider how meat must be seen as a part of the delicate natural balance to which all living things contribute.

# Appendix

**TABLE 1.** Red Meat and Poultry Consumption per Capita (lb/oz), World Regions, 1961

| *Region* | *Beef* | *Pork* | *Mutton and Goat* | *Poultry* | *Total* |
|---|---|---|---|---|---|
| World Average | 20/11 | 17/11 | 4/4 | 6/6 | 49/0 |
| Africa | 15/4 | 1/8 | 5/15 | 2/14 | 25/9 |
| Americas | 68/0 | 37/2 | 4/3 | 19/15 | 129/3 |
| Asia | 3/9 | 3/12 | 1/14 | 1/15 | 11/2 |
| Europe | 34/9 | 47/11 | 7/13 | 10/3 | 100/4 |
| Oceania | 88/14 | 24/2 | 93/12 | 9/4 | 215/15 |

*Source*: Data are compiled from FAOSTAT (Food and Agriculture Organization of the United Nations): Food Supply—Livestock and Fish Primary Equivalent, Food Supply Quantity, www.fao.org/faostat/en/#data/CL, unless otherwise noted.

*Note*: All data from FAOSTAT were converted from kg per capita to lb/oz per capita.

**TABLE 2.** Red Meat and Poultry Consumption per Capita (lb/oz), National Rankings, 1961

| *Country* | *Beef* | *Pork* | *Mutton and Goat* | *Poultry* | *Total* |
|---|---|---|---|---|---|
| Uruguay | 167/6 | 15/7 | 47/13 | 6/8 | 237/2 |
| New Zealand | 103/12 | 33/12 | 93/1 | 5/14 | 236/7 |
| Australia | 90/15 | 22/13 | 101/11 | 10/9 | 226/0 |
| Argentina | 184/0 | 19/3 | 15/5 | 4/8 | 223/0 |
| USA | 90/14 | 61/1 | 4/10 | 36/4 | 192/13 |
| Canada | 80/12 | 57/15 | 3/9 | 30/9 | 172/13 |
| United Kingdom | 54/13 | 55/12 | 25/6 | 13/14 | 149/13 |
| Germany | 43/15 | 82/3 | 1/10 | 11/0 | 138/12 |
| Brazil | 38/12 | 15/7 | 1/4 | 3/13 | 59/4 |
| Japan | 3/7 | 4/13 | 0/10 | 3/1 | 11/15 |
| China, mainland | 0/3 | 4/8 | 0/5 | 2/1 | 7/1 |
| 11-Country Average | 78/1 | 33/14 | 26/9 | 11/10 | 150/2 |

*Note*: These rankings include only nations focused on in this book.

**TABLE 3.** Red Meat and Poultry Consumption per Capita (lb/oz), World Regions, 2011

| *Region* | *Beef* | *Pork* | *Mutton and Goat* | *Poultry* | *Total* |
|---|---|---|---|---|---|
| World Average | 20/11 | 34/3 | 4/2 | 32/2 | 91/2 |
| Africa | 13/14 | 3/2 | 6/6 | 13/14 | 37/4 |
| Americas | 64/5 | 37/2 | 1/6 | 85/2 | 187/15 |
| Asia | 9/10 | 33/2 | 4/2 | 20/14 | 67/12 |
| Europe | 33/11 | 76/3 | 4/5 | 47/0 | 161/3 |
| Oceania | 87/11 | 47/3 | 22/10 | 93/0 | 250/8 |

**TABLE 4.** Red Meat and Poultry Consumption per Capita (lb/oz), National Rankings, 2011

| *Country* | *Beef* | *Pork* | *Mutton and Goat* | *Poultry* | *Total* |
|---|---|---|---|---|---|
| New Zealand | 104/0 | 42/13 | 45/3 | 77/14 | 269/14 |
| Australia | 91/3 | 50/14 | 20/0 | 99/13 | 261/14 |
| USA | 81/3 | 59/2 | 0/14 | 113/6 | 254/9 |
| Argentina | 121/2 | 19/14 | 3/13 | 84/13 | 229/10 |
| Brazil | 86/2 | 27/11 | 1/5 | 89/10 | 204/12 |
| Canada | 64/11 | 53/14 | 2/3 | 81/11 | 202/7 |
| Germany | 29/8 | 117/14 | 1/14 | 39/10 | 188/14 |
| United Kingdom | 41/6 | 58/13 | 10/5 | 67/14 | 178/6 |
| Uruguay | 69/14 | 28/3 | 21/6 | 46/3 | 165/10 |
| China, mainland | 10/6 | 78/8 | 6/8 | 27/5 | 122/11 |
| Japan | 19/13 | 45/3 | 0/5 | 42/0 | 107/5 |
| 11-Country Average | 65/6 | 53/0 | 10/5 | 70/0 | 198/11 |

**TABLE 5.** Grain and Meat Net Trade, World Regions, 1961 (export [+] and import [-] quantities in tons)

| *Region* | *Corn* | *Soybeans* | *Meat* |
|---|---|---|---|
| Africa | +1,204,165 | +4,433 | −23,344 |
| Americas | +8,400,199 | +3,454,811 | +133,727 |
| Asia | −2,273,927 | −1,125,537 | −88,613 |
| Europe | −7,586,437 | −2,250,248 | −778,868 |
| Oceania | −166 | −5 | +750,865 |

*Source*: The corn (maize) and soybean quantities are from FAOSTAT: Trade—Crops and Livestock Products, www.fao.org/faostat/en/#data/TP.

*Note*: Meat in tables 5 through 9 refers to the FAOSTAT aggregate trade in total meat, which includes beef, pork, mutton and goat, and poultry products, as well as duck, goose, game, rabbit, and marine mammals.

**TABLE 6.** Grain and Meat Net Trade, World Regions, 1973 (export [+] and import [-] quantities in tons)

| *Region* | *Corn* | *Soybeans* | *Meat* |
|---|---|---|---|
| Africa | +1,580,598 | −3,641 | +138,727 |
| Americas | +34,181,395 | +14,674,613 | −2,574 |
| Asia | −10,139,939 | −4,622,958 | −479,433 |
| Europe | −23,291,428 | −9,095,723 | −1,046,635 |
| Oceania | +17,140 | +1,173 | +1,661,327 |

**TABLE 7.** Grain and Meat Net Trade, World Regions, 1987 (export [+] and import [-] quantities in tons)

| *Region* | *Corn* | *Soybeans* | *Meat* |
|---|---|---|---|
| Africa | -1,428,946 | -95,160 | -470,216 |
| Americas | +38,168,854 | +24,752,417 | -218,326 |
| Asia | -25,966,138 | -7,530,878 | -1,551,612 |
| Europe | -10,842,855 | -17,329,693 | +891,353 |
| Oceania | +73,454 | -52 | +1,708,564 |

**TABLE 8.** Grain and Meat Net Trade, World Regions, 2000 (export [+] and import [-] quantities in tons)

| *Region* | *Corn* | *Soybeans* | *Meat* |
|---|---|---|---|
| Africa | -8,865,472 | -521,555 | -659,468 |
| Americas | +41,825,966 | +38,875,824 | +4,134,573 |
| Asia | -31,488,981 | -23,976,883 | -5,081,540 |
| Europe | -1,257,834 | -15,489,068 | +467,068 |
| Oceania | +37,534 | +7,017 | +2,264,698 |

**TABLE 9.** Grain and Meat Net Trade, World Regions, 2011 (export [+] and import [-] quantities in tons)

| *Region* | *Corn* | *Soybeans* | *Meat* |
|---|---|---|---|
| Africa | -11,030,002 | -2,175,470 | -1,937,418 |
| Americas | +49,520,353 | +82,453,517 | +11,421,040 |
| Asia | -42,577,307 | -67,907,180 | -9,105,022 |
| Europe | +5,686,653 | -12,694,181 | +720,213 |
| Oceania | +33,178 | -90 | +1,821,734 |

**TABLE 10.** Japan's Beef Production, 1960–2003

| *Year* | *Domestic Production (1,000 tons)* | *Imports (1,000 tons)* | *Exports (1,000 tons)* | *Supplies for Domestic Consumption (1,000 tons)* | *Self-Sufficiency Ratio* |
|---|---|---|---|---|---|
| 1960 | 141 | 6 | 0 | 147 | 96 |
| 1965 | 196 | 11 | 0 | 207 | 95 |
| 1970 | 282 | 33 | 0 | 315 | 90 |
| 1975 | 335 | 91 | 0 | 415 | 81 |
| 1980 | 431 | 172 | 0 | 597 | 72 |
| 1985 | 556 | 225 | 0 | 774 | 72 |
| 1990 | 555 | 549 | 0 | 1,095 | 51 |
| 1995 | 590 | 941 | 0 | 1,526 | 39 |
| 2000 | 521 | 1,055 | 0 | 1,554 | 34 |
| 2003 | 505 | 743 | 0 | 1,291 | 39 |

*Source*: *Historical Statistics of Japan*, Statistics Bureau, Ministry of Internal Affairs and Communications, chapter 7, "Agricultural, Forestry, and Fisheries," tables 7-55 and 7-56, www.stat.go.jp/english/data/chouki/index.htm.

**TABLE 11.** Japan's Pork Production, 1960–2003

| *Year* | *Domestic Production (1,000 tons)* | *Imports (1,000 tons)* | *Exports (1,000 tons)* | *Supplies for Domestic Consumption (1,000 tons)* | *Self-Sufficiency Ratio* |
|---|---|---|---|---|---|
| 1960 | 149 | 6 | 0 | 155 | 96 |
| 1965 | 431 | 0 | 0 | 431 | 100 |
| 1970 | 779 | 17 | 0 | 796 | 98 |
| 1975 | 1,023 | 208 | 0 | 1,190 | 86 |
| 1980 | 1,430 | 207 | 0 | 1,646 | 87 |
| 1985 | 1,559 | 272 | 0 | 1,813 | 86 |
| 1990 | 1,536 | 488 | 0 | 2,066 | 77 |
| 1995 | 1,299 | 772 | 0 | 2,095 | 62 |
| 2000 | 1,256 | 952 | 0 | 2,188 | 57 |
| 2003 | 1,274 | 1,145 | 0 | 2,406 | 53 |

*Source: Historical Statistics of Japan*, Statistics Bureau, Ministry of Internal Affairs and Communications, chapter 7, "Agricultural, Forestry, and Fisheries," tables 7-55 and 7-56, www.stat.go.jp/english/data/chouki/index.htm.

**TABLE 12.** Japan's Chicken Production, 1960–2003

| *Year* | *Domestic Production (1,000 tons)* | *Imports (1,000 tons)* | *Exports (1,000 tons)* | *Supplies for Domestic Consumption (1,000 tons)* | *Self-Sufficiency Ratio* |
|---|---|---|---|---|---|
| 1960 | 103 | 0 | 0 | 103 | 100 |
| 1965 | 238 | 8 | 0 | 246 | 97 |
| 1970 | 496 | 12 | 1 | 507 | 98 |
| 1975 | 759 | 28 | 3 | 784 | 97 |
| 1980 | 1,120 | 80 | 4 | 1,194 | 94 |
| 1985 | 1,354 | 115 | 3 | 1,466 | 92 |
| 1990 | 1,380 | 297 | 8 | 1,678 | 82 |
| 1995 | 1,252 | 581 | 3 | 1,820 | 69 |
| 2000 | 1,195 | 686 | 3 | 1,865 | 64 |
| 2003 | 1,239 | 585 | 3 | 1,848 | 67 |

*Source*: *Historical Statistics of Japan*, Statistics Bureau, Ministry of Internal Affairs and Communications, chapter 7, "Agricultural, Forestry, and Fisheries," tables 7-55 and 7-56, www.stat.go.jp/english/data/chouki/index.htm.

**TABLE 13.** Japan's Red Meat and Poultry Consumption, 1929–2011 (lbs/oz per capita per year)

| *Year* | *Average Amount* | *World Average Amount* |
|---|---|---|
| 1929 | 2/8[a] | N/A |
| 1934–1938 | 4/15[b] | N/A |
| 1948–1950 | 4/1[c] | N/A |
| 1961 | 16/11[d] | 48/15[e] |
| 1965 | 20/0 | 53/13 |
| 1970 | 34/13 | 57/11 |
| 1975 | 48/5 | 60/13 |
| 1980 | 64/6 | 65/12 |
| 1985 | 72/10 | 67/13 |
| 1990 | 83/11 | 72/4 |
| 1995 | 96/8 | 75/10 |
| 2000 | 99/6 | 80/8 |
| 2005 | 102/9 | 84/2 |
| 2011 | 107/5 | 91/3 |

[a]Taiguin I., "Japan and Overpopulation," *Pacific Affairs* 2, no. 7 (July 1929): 407.

[b]Ishige Naomichi, *The History and Culture of Japanese Food* (London: Kegan Paul, 2001), 153.

[c]Dale E. Hathaway, Hendrik S. Houthakker, and John A. Schnittker, "Food Prices and Inflation," *Brookings Papers on Economic Activity* 1974, no. 1 (1974): 91.

[d]1961–2011 Japan data from FAOSTAT: Food Supply—Livestock and Fish Primary Equivalent, Food Supply Quantity, www.fao.org/faostat/en/#data/CL

[e]1961–2011 world average data from FAOSTAT: Food Supply—Livestock and Fish Primary Equivalent, Food Supply Quantity, www.fao.org/faostat/en/#data/CL

# Notes

## Introduction

1. Notable literature highlighting meat's negative consequences during the contemporary era include Jeremy Rifkin, *Beyond Beef: The Rise and Fall of the Cattle Culture* (New York: Dutton, 1992); Ken Midkiff, *The Meat You Eat: How Corporate Farming Has Endangered America's Food Supply* (New York: St. Martin's, 2004); Paul Roberts, *The End of Food* (Boston: Houghton Mifflin, 2008); Tony Weis, *The Ecological Hoofprint: The Global Burden of Industrial Livestock* (London: Zed Books, 2013); Ted Genoways, *The Chain: Farm, Factory, and the Fate of Our Food* (New York: HarperCollins, 2014); James McWilliams, *The Modern Savage: Our Unthinking Decision to Eat Animals* (New York: St. Martin's, 2015); Marta Zaraska, *Meathooked: The History and Science of Our 2.5-Million-Year Obsession with Meat* (New York: Basic Books, 2016); and Ellen K. Silbergeld, *Chickenizing Farms and Food: How Industrial Meat Production Endangers Workers, Animals, and Consumers* (Baltimore: Johns Hopkins University Press, 2016). Anti-meat films include *Earthlings* (2005); *Food, Inc.* (2008); *Vegucated* (2011); *Forks over Knives* (2011); and *Cowspiracy: The Sustainability Secret* (2014).

2. On early humans' meat-eating proclivities, see Richard Wrangham, *Catching Fire: How Cooking Made Us Human* (New York: Basic Books, 2009); B. W. Higman, *How Food Made History* (Chichester, West Sussex, UK: Wiley-Blackwell, 2012); Vaclav Smil, *Should We Eat Meat? Evolution and Consequences of Modern Carnivory* (Chichester, West Sussex, UK: Wiley-Blackwell, 2013); Stephen Le, *100 Million Years of Food: What Our Ancestors Ate and Why It Matters Today* (New York: Picador, 2016), 43–60; and Zaraska, *Meathooked*, 168–170. On Japan's dietary transition, see Vaclav Smil and Kobayashi Kazuhiko, *Japan's Dietary Transition and Its Impacts* (Cambridge: MIT Press, 2012). A popular treatment of China's dietary transition is Paul French and Matthew Crabbe, *Fat China: How Expanding Waistlines Are Changing a Nation* (London: Anthem, 2010).

3. Merrill Kelley Bennett, "Wheat in National Diets," *Wheat Studies of the Food Research Institute* 18, no. 2 (1941), 37–76. On Bennett's Law, also see Jeffery Sobal, "Food System Globalization, Eating Transformations, and Nutrition Transitions," in *Food in Global History*, ed. Raymond Grew (Boulder, CO: Westview, 1999), 171–193. For a brief sampling of studies that rely implicitly on Bennett's Law, see Eric B. Ross, "An Overview of Trends in Dietary Variation from Hunter-Gatherer to Modern Capitalist Societies," in *Food and Evolution: Toward a Theory of Human Food Habits*, ed. Marvin Harris and Eric B. Ross (Philadelphia: Temple University Press, 1987), 27–42; David Grigg, "The Changing Geography of World Food Consumption in the Second Half of the Twentieth Century," *Geographical Journal* 165, no. 1 (1999): 1–11; Derek J. Oddy, *From Plain Fare to Fusion Food: British Diet from the 1890s to the 1990s* (Suffolk, UK: Boydell, 2003);

Hans Jürgen Teuteberg, "Urbanization and Nutrition: Historical Research Reconsidered," in *Food and the City in Europe since 1800*, ed. Peter J. Atkins, Peter Lummel, and Derek J. Oddy (Aldershot, UK: Ashgate, 2007), 13–23; Roser Nicolau-Nos and Josep Pujol-Andreu, "Urbanization and Dietary Change in Mediterranean Europe: Barcelona, 1870–1935," in *Food and the City in Europe since 1800*, ed. Peter J. Atkins, Peter Lummel, and Derek J. Oddy (Aldershot, UK: Ashgate, 2007), 39–49; and Peter Dauvergne, *The Shadows of Consumption: Consequences for the Global Environment* (Cambridge: MIT Press, 2008).

4. Two of the most important articles on food regimes theory are Harriet Friedmann, "The Political Economy of Food: The Rise and Fall of the Postwar International Food Order," *American Journal of Sociology* 88, Supplement (1982): S248–S286; and Harriet Friedmann and Philip McMichael, "Agriculture and the State System: The Rise and Decline of National Agricultures, 1870 to the Present," *Sociologia Ruralis* 29, no. 2 (1989): 93–117.

5. Jean Anthelme Brillat-Savarin, *The Physiology of Taste: or, Meditations on Transcendental Gastronomy*, trans. M. F. K. Fisher (New York: Heritage, 1949; originally published in 1825), 91, 93.

6. On the status of fish in traditional English and French diets, see Stephen Mennell, *All Manners of Food: Eating and Taste in England and France from the Middle Ages to the Present*, 2nd ed. (Urbana: University of Illinois Press, 1996); Ann Hagen, *Anglo-Saxon Food and Drink: Production, Processing, Distribution and Consumption* (Norfolk, UK: Anglo-Saxon Books, 2006); and Joan Thirsk, *Food in Early Modern England: Phases, Fads, Fashions, 1500–1760* (London: Hambledon Continuum, 2007). On the issue of Buddhism and fish consumption in Japan, see Frederick J. Simoons, *Eat Not This Flesh: Food Avoidances from Prehistory to the Present*, 2nd ed., revised and enlarged (Madison: University of Wisconsin Press, 1994), 288–294. On the history of dairy production and consumption, see Deborah Valenze, *Milk: A Local and Global History* (New Haven, CT: Yale University Press, 2011); and Kendra Smith-Howard, *Pure and Modern Milk: An Environmental History since 1900* (New York: Oxford University Press, 2013). On the rise of dairy consumption in India, see Gordon R. Hopper, "Changing Food Production and Quality of Diet in India, 1947–98," *Population and Development Review* 25, no. 3 (1999): 443–477; and Valenze, *Milk*, 270–278.

7. Richard Perren, *Taste, Trade and Technology: The Development of the International Meat Industry since 1840* (Aldershot, UK: Ashgate, 2006); and Smil, *Should We Eat Meat?* Two other books provide important overarching cultural perspectives on the ways in which meat has impacted Western intellectual and cultural history. Carol Adams's *The Sexual Politics of Meat: A Feminist-Vegetarian Critical Theory* suggests how killing animals for meat has justified patriarchy and violence in the West. Likewise, Nick Fiddes's *Meat: A Natural Symbol* critiques the powerful cultural significance of meat in Western society. Both books have been influential in shaping the way that contemporary literature has portrayed, in particular, meat's harmful effects around the world. While Adams's and Fiddes's views are certainly insightful and have been widely adopted in perspectives on meat's importance, they downplay how people around the globe, especially outside the West, not only wanted meat as part of their diet but also wanted to identify with the power that eating meat signified. See Carol Adams, *The Sexual Politics of Meat: A Feminist-Vegetarian Critical Theory* (New York: Continuum, 1996; originally published in 1990); and Nick Fiddes, *Meat: A Natural Symbol* (London: Routledge, 1991).

8. Sidney W. Mintz, *Sweetness and Power: The Place of Sugar in Modern History* (New York: Penguin Books, 1985).

## Chapter 1

1. On the first food regime, see Friedmann, "The Political Economy of Food: The Rise and Fall," S256–S257; Friedmann and McMichael, "Agriculture and the State System," 93–117; and Philip D. McMichael, "Tensions between National and International Control of the World Food Order: Contours of a New Food Regime," *Sociological Perspectives* 35, no. 2 (1992): 345. The first food regime was partly dependent on important changes in European agriculture during the same period. See J. L. Van Zanden, "The First Green Revolution: The Growth of Production and Productivity in European Agriculture, 1870–1914," *Economic History Review* 44, no. 2 (1991): 215–239.

2. Jeffrey M. Pilcher, *Food in World History* (New York: Routledge, 2006), 55.

3. Vincent J. Knapp, "The Democratization of Meat and Protein in Late Eighteenth- and Nineteenth-Century Europe," *Historian* 59, no. 3 (1997): 542–543.

4. Vaclav Smil, "Eating Meat: Evolution, Patterns, and Consequences," *Population and Development Review* 28, no. 4 (2002): 607; Christopher Dyer, "Did the Peasants Really Starve in Medieval England?," in *Food and Eating in Medieval Europe*, ed. Martha Carlin and Joel T. Rosenthal (London: Hambledon, 1998), 58–59; Margaret Murphy, "Feeding Medieval Cities: Some Historical Approaches," in *Food and Eating in Medieval Europe*, ed. Martha Carlin and Joel T. Rosenthal (London: Hambledon, 1998), 120; Esther Pascua, "From Forest to Farm and Town: Domestic Animals from ca. 1000 to ca. 1450," in *A Cultural History of Animals in the Medieval Age*, ed. Brigitte Resl (Oxford, UK: Berg, 2007), 98; Christopher Dyer, "Changes in Diet in the Late Middle Ages: The Case of Harvest Workers," in *Everyday Life in Medieval England*, ed. Christopher Dyer (London: Hambledon, 1994), 85–86; Catharina Lis and Hugo Soly, *Poverty and Capitalism in Pre-Industrial Europe* (Atlantic Highlands, NJ: Humanities, 1979), 74–75, 113, 178; James C. Riley, "A Widening Market in Consumer Goods," in *Early Modern Europe: An Oxford History*, ed. Euan Cameron (Oxford: Oxford University Press, 1999), 243; Mennell, *All Manners of Food*, 44; Hans Jürgen Teuteberg, "The General Relationship between Diet and Industrialization," in *European Diet from Pre-Industrial to Modern Times*, ed. Elborg Forster and Robert Forster (New York: Harper Torchbooks, 1975), 61–109; and Ken Albala, *Food in Early Modern Europe* (Westport, CT: Greenwood, 2003), 62.

5. Elaine N. McIntosh, *American Food Habits in Historical Perspective* (Westport, CT: Praeger, 1995), 82; and Evan Jones, *American Food: The Gastronomic Story*, 3rd ed. (Woodstock, NY: Overlook, 2007), 17–18.

6. Sarah F. McMahon, "'All Things in Their Proper Season': Seasonal Rhythms of Diet in Nineteenth Century New England," *Agricultural History* 63, no. 2 (1989): 132.

7. Andrew Sluyter, *Black Ranching Frontiers: African Cattle Herders of the Atlantic World, 1500–1900* (New Haven: Yale University Press, 2012), 169–217.

8. Terry G. Jordan, *North American Cattle-Ranching Frontiers: Origins, Diffusion, and Differentiation* (Albuquerque: University of New Mexico Press, 1993), 34–35; Simon M. Evans, "Stocking the Canadian Range," *Alberta History* 26, no. 3 (1978): 1–8; Evans, "The Origin of Ranching in Western Canada: American Diffusion or Victorian Transplant?," *Great Plains Quarterly* 3, no. 2 (1983): 79–91; Lawrence M. Woods, *British Gentlemen in the Wild West: The Era of the Intensely English Cowboy* (London: Robson, 2003; originally published in 1990); Perren, *Taste, Trade and Technology*, 31, 34–37; and Dorothee Brantz," The Domestication of Empire: Human-Animal Relations at the Intersection of Civilization, Evolution, and Acclimatization in the Nineteenth Century," in *A Cultural History of Animals in the Age of Empire*, ed. Kathleen Kete (Oxford, UK: Berg, 2007), 85–86. On the early history of livestock domestication in Europe, see Peter

Edwards, "Domesticated Animals in Renaissance Europe," in *A Cultural History of Animals in the Renaissance*, ed. Bruce Boehner (Oxford, UK: Berg, 2007), 84–85; Alistair Hennessy, *The Frontier in Latin American History* (Albuquerque: University of New Mexico Press, 1978), 82–89; A. L. Lloyd, "Meat from Argentina: The History of a National Industry," *History Today* 1, no. 3 (1951): 32–34; and Sam White, "From Globalized Pig Breeds to Capitalist Pigs: A Study in Animal Cultures and Evolutionary History," *Environmental History* 16 (January 2011): 94–120.

9. On humans' natural taste for meat flavor, or umami, which is now considered one of the five basic tastes (with sweetness, sourness, bitterness, and saltiness), see Zaraska, *Meathooked*, 59–68; and Fiddes, *Meat*, 30–54.

10. Rebecca Earle, *The Body of the Conquistador: Food, Race and the Colonial Experience in Spanish America, 1492–1700* (Cambridge: Cambridge University Press, 2012), 55. Earle describes Spanish dietary encounters in the New World.

11. On the English demand for beef, see Ben Rogers, *Beef and Liberty: Roast Beef, John Bull and the English Nation* (London: Vintage, 2004), 75–94; Craig Mulgrew, *Food, Energy and the Creation of Industriousness: Work and Material Culture in Agrarian England, 1550–1780* (Cambridge: Cambridge University Press, 2011), 85–86; and Menno Spiering, "Food, Phagophobia and English National Identity," *European Studies* 22 (2006): 31–48.

12. Song lyrics quoted in Spiering, "Food, Phagophobia and English National Identity," 37.

13. Peter Kalm, quoted in Karen Raber, "From Sheep to Meat, from Pets to People: Animal Domestication, 1600–1800," in *A Cultural History of Animals in the Age of Enlightenment*, ed. Matthew Senior (Oxford, UK: Berg, 2007), 83.

14. On eighteenth-century French attitudes about meat, see Sydney Watts, *Meat Matters: Butchers, Politics, and Market Culture in Eighteenth-Century Paris* (Rochester, NY: University of Rochester Press, 2006), 2.

15. Sam White, "From Globalized Pig Breeds to Capitalist Pigs," 106; K. Grady, "The Cattle and Meat Trades in Leeds, 1780–1900," *Northern History* 37 (December 2000): 140–145; Laura Mason, *Food Culture in Great Britain* (Westport, CT: Greenwood, 2004), 25–27; Steven Shapin, "'You Are What You Eat': Historical Changes in Ideas about Food and Identity," *Historical Research* 87, no. 237 (2014): 384–390; and Friedrich Engels, *The Condition of the Working Class in England*, trans. and ed. W. O. Henderson and W. H. Chaloner (Stanford, CA: Stanford University Press, 1990; originally published in 1845), 85. On German meat preferences, see Ursula Heinzelmann, *Food Culture in Germany* (Westport, CT: Greenwood, 2008), 16–17, 49, 86; and Nika Standen Hazelton, *The Cooking of Germany* (New York: Time-Life Books, 1969), 12, 35–36.

16. Roger Horowitz, Jeffrey M. Pilcher, and Sydney Watts, "Meat for the Multitudes: Market Culture in Paris, New York City, and Mexico City over the Long Nineteenth Century," *American Historical Review* 109, no. 4 (2004): 1059; and Sydney Watts, "The Grand Boucherie, the 'Right' to Meat, the Growth of Paris," in *Meat, Modernity, and the Rise of the Slaughterhouse*, ed. Paula Young Lee (Lebanon: University of New Hampshire Press, 2008), 24. On the significant effects that changing agricultural emphases and new railway connections between urban and rural areas had on increasing meat consumption, see Nicolau-Nos and Pujol-Andreu, "Urbanization and Dietary Change," 39–49.

17. Peter J. Atkins, "'A Tale of Two Cities': A Comparison of Food Supply in London and Paris in the 1850s," in *Food and the City in Europe since 1800*, ed. Peter J. Atkins, Peter Lummel, and Derek J. Oddy (Aldershot, UK: Ashgate, 2007), 25; Teuteberg, "Ur-

banization and Nutrition," 15–20; Horowitz, Pilcher, and Watts, "Meat for the Multitudes," 1055–1082; Sydney Watts, *Meat Matters*, 8–14; George Dodd, *The Food of London* (New York: Arno, 1976; originally published in 1856), 227, 272, 282, 325; and David P. Nally, *Human Encumbrances: Political Violence and the Great Irish Famine* (Notre Dame, IN: University of Notre Dame Press, 2011), 42, 205. On the increase in meat consumption among working-class Parisians in the mid- to late-nineteenth century, see Anne Lhuissier, "Cuts and Classification: The Use of Nomenclatures as a Tool for the Reform of the Meat Trade in France, 1850–1880," *Food and Foodways* 10 (2003): 183–208.

18. Sydney Watts, "The Grand Boucherie," 26; Paula Young Lee, "Siting the Slaughterhouse: From Shed to Factory," in *Meat, Modernity, and the Rise of the Slaughterhouse*, ed. Paula Young Lee (Lebanon: University of New Hampshire Press, 2008), 52–62; Dorothee Brantz, "Stunning Bodies: Animal Slaughter, Judaism, and the Meaning of Humanity in Imperial Germany," *Central European History* 35, no. 2 (2002): 169; Brantz, "Animal Bodies, Human Health, and the Reform of Slaughterhouses in Nineteenth-Century Berlin," *Food and History* 3, no. 2 (2005): 194, 213–214; Kyri Claflin, "La Villette: City of Blood (1867–1914)," in *Meat, Modernity, and the Rise of the Slaughterhouse*, ed. Paula Young Lee (Lebanon: University of New Hampshire Press, 2008), 27–28; Chris Otter, "Civilizing Slaughter: The Development of the British Public Abattoir, 1850–1910," in *Meat, Modernity, and the Rise of the Slaughterhouse*, ed. Paula Young Lee (Lebanon: University of New Hampshire Press, 2008), 89–106; and Grady, "The Cattle and Meat Trades in Leeds," 133–155.

19. Roger Horowitz, *Putting Meat on the American Table: Taste, Technology, Transformation* (Baltimore: Johns Hopkins University Press, 2006), 20–21; Margaret Walsh, *The Rise of the Midwestern Meat Packing Industry* (Lexington: University Press of Kentucky, 1982), 1–37, 89–92, 94; Evan Jones, *American Food*, 17–21; Sam Bowers Hilliard, *Hog Meat and Hoecake: Food Supply in the Old South, 1840–1860* (Carbondale: Southern Illinois University Press, 1972), 68, 105; Rudolf Alexander Clemen, *The American Livestock and Meat Industry* (New York: Ronald Press, 1923), 303–305; and Wilson J. Warren, *Tied to the Great Packing Machine: The Midwest and Meatpacking* (Iowa City: University of Iowa Press, 2007), 7–10, 145, 156.

20. Horowitz, *Putting Meat on the American Table*, 29–30; and Warren, *Tied to the Great Packing Machine*, 15–16.

21. Simon G. Hanson, *Argentine Meat and the British Market: Chapters in the History of the Argentine Meat Industry* (Stanford, CA: Stanford University Press, 1938), 3–9, 16–17, 29–32, 104; Perren, *Taste, Trade and Technology*, 45–46; Andrew Graham-Yooll, *The Forgotten Colony: A History of the English-Speaking Communities in Argentina* (London: Hutchinson, 1981), 227; Jack Colin Crossley, "The Location and Development of the Agricultural and Industrial Enterprises of Liebig's Extract of Meat Company in the River Plate Countries," (PhD diss., University of Leicester, 1973), 1–2, 54–111; and Clarence F. Jones, "The Trade of Uruguay," *Economic Geography* 3, no. 3 (1927), 366.

22. Derek J. Oddy, "From Roast Beef to Chicken Nuggets: How Technology Changed Meat Consumption in Britain in the Twentieth Century," in *The Food Industries of Europe in the Nineteenth and Twentieth Centuries*, ed. Derek J. Oddy and Alain Drouard (Farnham, Surrey, UK: Ashgate, 2013), 236; Mason, *Food Culture in Great Britain*, 48; Richard Perren, "The Meat and Livestock Trade in Britain, 1850–70," *Economic History Review* 28, no. 3 (1975): 388; and Ian MacLachlan, "A Bloody Offal Nuisance: The Persistence of Private Slaughter-Houses in Nineteenth-Century London," *Urban History* 34, no. 2 (2007): 228–234.

23. Perren, *Taste, Trade and Technology*, 9–10; Simon M. Evans, "Canadian Beef for Victorian Britain," *Agricultural History* 53, no. 4 (1979): 749; Friedmann and McMichael, "Agriculture and the State System," 93–117; Richard Perren, "Big Business and Its Customers: The European Market for American Meat from 1840 to 1939," *Journal of European Economic History* 32, no. 3 (2003): 592–597; John P. Hullman, "British Meat Imports in the Free Trade Era," *Agricultural History* 52, no. 2 (1978): 253–254; and Walsh, *The Rise of the Midwestern Meat Packing Industry*, 36.

24. Pilcher, *Food in World History*, 57; John Burnett, *Plenty and Want: A Social History of Diet in England from 1815 to the Present Day*, rev. ed. (London: Scolar, 1979), 137; William David Zimmerman, "Live Cattle Export Trade between United States and Great Britain, 1868–1885," *Agricultural History* 36, no. 1 (1962): 46–52; Richard Perren, "The North American Beef and Cattle Trade with Great Britain, 1870–1914," *Economic History Review* 24, no. 3 (1971): 430–444; C. Knick Harley, "Steers Afloat: The North American Meat Trade, Liner Predominance, and Freight Rates, 1870–1913," *Journal of Economic History* 68, no. 4 (2008): 1028–1037; Evans, "Canadian Beef for Victorian Britain," 749–755; and Hullman, "British Meat Imports in the Free Trade Era," 252–253. On the effect of Irish exports of meat and other foods to Britain, see Hasia R. Diner, *Hungering for America: Italian, Irish, and Jewish Foodways in the Age of Migration* (Cambridge, MA: Harvard University Press, 2001), 84–112.

25. John L. Gignilliat, "Pigs, Politics, and Protection: The European Boycott of American Pork, 1879–1891," *Agricultural History* 35, no. 1 (1961): 4; James C. Hunt, "Peasants, Grain Tariffs, and Meat Quotas: Imperial German Protectionism Reexamined," *Central European History* 7, no. 4 (1974): 314–318; and Brantz, "The Domestication of Empire," 83. Hunt notes that before the pork war heated up in the mid-1880s, Germany's imports of live cattle averaged between 4 and 8 percent of total weight of slaughtered cattle. Imports of live hogs averaged between 2 and 6 percent between 1889 and 1894. Live hog imports continued to fall after 1894 and averaged only 0.3 percent after 1904 through World War I.

26. Ian Arthur, "Shipboard Refrigeration and the Beginning of the Frozen Meat Trade," *Journal of the Royal Australian Historical Society* 92, no. 1 (2006): 63–82; Peter H. Smith, *Politics and Beef in Argentina: Patterns of Conflict and Change* (New York: Columbia University Press, 1969, 32–53; Clarence F. Jones, "Argentine Trade Developments," *Economic Geography* 2, no. 3 (1926): 365–370; Clarence F. Jones, "The Trade of Uruguay," 366–369; S. B. Saul, "Britain and World Trade, 1870–1914," *Economic History Review* 7, no. 1 (1954): 63–64; Hullman, "British Meat Imports in the Free Trade Era," 254–255; and J. D. Whelpley, "American Control of England's Food Supply," *North American Review* 174, no. 547 (June 1902): 801.

27. Ian MacLachlan, *Kill and Chill: Restructuring Canada's Beef Commodity Chain* (Toronto: University of Toronto Press, 2001), 123–124, 144–145; H. E. Bronson, "The Saskatchewan Meat Packing Industry: Some Historical Highlights," *Saskatchewan History* 26, no. 1 (1973): 24–37; and Jim Silver, "The Origins of Winnipeg's Packinghouse Industry: Transitions from Trade to Manufacture," *Prairie Forum* 19, no. 1 (1994): 15–30.

28. Perren, *Taste, Trade and Technology*, 42–44; Keith Farrer, *To Feed a Nation: A History of Australian Food Science and Technology* (Collingwood, AU: CSIRO Publishing, 2005), 34–47; John Louis William Thudichum, *On the Origin, Nature, and Uses of Liebig's Extract; With an Analytical Comparison of Other Essences and Preparations of Meat* (London: John Churchill and Sons, 1869), 6; and G. J. R. Linge, *Industrial Awakening: A Geography of Australian Manufacturing, 1788 to 1890* (Canberra: Australian National University Press, 1979), 10, 240–241, 451.

29. Perren, *Taste, Trade and Technology*, 48–49; Linge, *Industrial Awakening*, 242, 453–454; and Derek J. Oddy, "The Growth of Britain's Refrigerated Meat Trade, 1880–1939," *Mariner's Mirror* 93, no. 3 (2007): 269–280.

30. Ian Hunter, "Commodity Chains and Networks in Emerging Markets: New Zealand, 1880–1910," *Business History Review* 79, no. 2 (2005): 275–304; Perren, *Taste, Trade and Technology*, 61; Michael M. Roche, "International Food Regimes: New Zealand's Place in the International Frozen Meat Trade, 1870–1935," *Historical Geography* 27 (1999): 138; Philip S. E. Hereford, *The New Zealand Frozen Meat Trade* (Wellington: New Zealand Publishing, 1932); and Bill Maughan, "Red Meat," in *The Structure and Dynamics of New Zealand Industries*, ed. Michael Pickford and Allan Bollard, 27 (Palmerston, NZ: Dunmore, 1998).

31. Robert Greenhill, "Shipping and the Refrigerated Meat Trade from the River Plate, 1900–1930," *International Journal of Maritime History* 4, no. 1 (1992): 69–70; Perren, *Taste, Trade and Technology*, 32–33, 49; Clarence F. Jones, "Argentine Trade Developments," 365–366; Hanson, *Argentine Meat and the British Market*, 48–97, 99–101, 107–118, 187; A. J. Christopher, "Patterns of British Overseas Investment in Land, 1885–1913," *Transactions of the Institute of British Geographers* 10, no. 4 (1985): 461–462; Crossley, "Location and Development," 116–117, 137–141; Peter H. Smith, *Politics and Beef in Argentina*, 32–34; and Richard Perren, "Farmers and Consumers under Strain: Allied Meat Supplies in the First World War," *Agricultural History Review* 53, no. 2 (2005): 212–228.

32. Perren, *Taste, Trade and Technology*, 24–25, 60–61; and Linge, *Industrial Awakening*, 15.

33. Yves Guyot, "The Bread and Meat of the World," *Publications of the American Statistical Association* 9, no. 67/68 (September-December 1904): 111. Hereford's estimate of 2.5 pounds of meat consumed per week circa 1900 in New Zealand is found in *The New Zealand Frozen Meat Trade*, 38.

34. McIntosh, *American Food Habits*, 82, 216; Horowitz, *Putting Meat on the American Table*, 11–12, 32, 41; Richard Osborn Cummings, *The American and His Food* (New York: Arno Press and New York Times, 1970; originally published in 1941), appendix B; Eric B. Ross, "Patterns of Diet and Forces of Production: An Economic and Ecological History of the Ascendancy of Beef in the United States Diet," in *Beyond the Myths of Culture: Essays in Cultural Materialism*, ed. Eric B. Ross (London: Academic Press, 1980)," 190–204; Warren, *Tied to the Great Packing Machine*, 221–223; Guyot, "The Bread and Meat of the World," 111; and USDA, BAI, *Twenty-Sixth Annual Report for the Year 1909* (Washington, DC: GPO, 1911), 315.

35. Fitzgibbon, "Uruguay's Agricultural Problems," 252–253; Hilda Sabato, *Agrarian Capitalism and the World Market: Buenos Aires in the Pastoral Age, 1840–1890* (Albuquerque: University of New Mexico Press, 1990), 23, 31; Greenhill, "Shipping and the Refrigerated Meat Trade," 68; Stephen Bell, "Early Industrialization in the South Atlantic: Political Influences on the *Charqueadas* of Rio Grande do Sul before 1860," *Journal of Historical Geography* 19, no. 4 (1993): 399; Hanson, *Argentine Meat and the British Market*, 7–9, 27–29, 102–104; and Crossley, "Location and Development," 116, 140–142.

36. Keir Waddington, *The Bovine Scourge: Meat, Tuberculosis and Public Health, 1850–1914* (Woodbridge, UK: Boydell, 2006), 15; Oddy, "From Roast Beef to Chicken Nuggets," 236; Burnett, *Plenty and Want*, 131; Hullman, "British Meat Imports in the Free Trade Era," 250–251; Reay Tannahill, *Food in History* (New York: Stein and Day, 1973), 321; Van Zanden, "The First Green Revolution," 227; Knapp, "Democratization of Meat and Protein," 549–550; Perren, "Big Business and Its Customers," 603; and United States

Department of Agriculture (USDA), Bureau of Animal Industry (BAI), *Twenty-Second Annual Report for the Year 1905* (Washington, DC: GPO, 1907), 284. Guyot, "The Bread and Meat of the World," 111, estimated that the average meat consumption in the United Kingdom from 1891 to 1900 was 122 pounds. The meat consumption gap between affluent craftsmen and poor workers in Britain closed between 1904 and 1918. See Ian Gazeley and Andrew Newell, "The First World War and Working-Class Food Consumption in Britain," *European Review of Economic History* 17 (2013): 71–94. On the gap between meat consumption between France's rich and poor in the late nineteenth century, see Georges Vigarello, *The Metamorphoses of Fat: A History of Obesity*, trans. C. Jon Delogu (New York: Columbia University Press, 2013), 152.

37. Tannahill, *Food in History*, 321.

38. Wilhelm Abel, *Agricultural Fluctuations in Europe: From the Thirteenth to the Twentieth Centuries*, trans. Olive Ordish (New York: St. Martin's, 1980; originally published in 1966), 267.

39. Lizzie Collingham, *The Taste of War: World War II and the Battle for Food* (New York: Penguin, 2012), 18.

40. Ursula Heinzelmann, *Food Culture in Germany*, 16–17, 49, 86; Nika Standen Hazelton, *The Cooking of Germany*, 12, 35–36; USDA, BAI, *Twenty-Second Annual Report for the Year 1905*, 284–285; USDA, BAI, *Twenty-Sixth Annual Report for the Year 1909*, 314–316; Guyot, "The Bread and Meat of the World," 111; Steven B. Webb, "Agricultural Protection in Wilhelminian Germany: Forging an Empire with Pork and Rye," *Journal of Economic History* 42, no. 2 (1982): 310, 317; and Teuteberg, "Urbanization and Nutrition," 17–19. The USDA reported in 1911 that Germany was the only country in Europe that officially recognized the slaughter of dogs for human consumption.

## Chapter 2

1. Rui Wang, "Fishing, Farming, and Animal Husbandry in the Early and Middle Neolithic of the Middle Yellow River Valley China" (PhD diss., University of Illinois at Urbana-Champaign, 2004); Jacqueline M. Newman, *Food Culture in China* (Westport, CT: Greenwood, 2004), 2; Eugene N. Anderson, *The Food of China* (New Haven, CT: Yale University Press, 1988), 24, 51; Frederick J. Simoons, *Food in China: A Cultural and Historical Inquiry* (Boca Raton, FL: CRC Press, 1991), 32–35; Roel Sterckx, "Food and Philosophy in Early China," in *Of Tripod and Palate: Food, Politics, and Religion in Traditional China*, ed. Roel Sterckx (New York: Palgrave Macmillan, 2005), 38–52.

2. Roel Sterckx, *Food, Sacrifice, and Sagehood in Early China* (Cambridge: Cambridge University Press, 2011), 30; see also 11–30.

3. John Kieschnick, "Buddhist Vegetarianism in China," in *Of Tripod and Palate: Food, Politics, and Religion in Traditional China*, ed. Roel Sterckx (New York: Palgrave Macmillan, 2005), 186–206.

4. Kitaguchi Suehiro, *An Introduction to the Buraku Issue: Questions and Answers*, trans. Alastair McLauchlan (Richmond, Surrey, UK: Japan Library, 1999; originally published 1986–1993); Richard M. Jaffe, *Neither Monk nor Layman: Clerical Marriage in Modern Japanese Buddhism* (Honolulu: University of Hawai'i Press, 2001), 29; Hans Martin Krämer, "'Not Befitting Our Divine Country': Eating Meat in Japanese Discourses of Self and Other from the Seventeenth Century to the Present," *Food and Foodways* 16, no. 1 (2008): 36–37; Ishige Naomichi, *The History and Culture of Japanese Food* (London: Kegan Paul, 2001), 53, 57; Smil and Kobayashi, *Japan's Dietary Transition*

*and Its Impacts*, 10–11, 48; and Smil, *Should We Eat Meat?*, 64. On the continuation of emperors eating wild game until the 12th century, see Pieter S. de Ganon, "The Animal Economy" (PhD diss., Princeton University, 2011), 135. Katarzyna J. Cwiertka in *Modern Japanese Cuisine: Food, Power, and National Identity* (London: Reaktion Books, 2006), 24–25, argues that Buddhist concerns had little to do with the ban and that it was only weakly enforced until the sixteenth century. She says the original intent of the ban was to protect limited livestock resources that were needed for traction power. Ishige notes a similar impetus for the meat ban (*History and Culture of Japanese Food*, 55).

5. Vincent Goossaert, "The Beef Taboo and the Sacrificial Structure of Late Imperial Chinese Society," in *Of Tripod and Palate: Food, Politics, and Religion in Traditional China*, ed. Roel Sterckx (New York: Palgrave Macmillan, 2005), 237–246; Eugene N. Anderson, *The Food of China*, 69–86, 129; Michael Freeman, "Sung," in *Food in Chinese Culture: Anthropological and Historical Perspectives*, ed. K. C. Chang (New Haven, CT: Yale University Press 1977), 144–164; Newman, *Food Culture in China*, 2, 11, 59; Simoons, *Food in China*, 31–37.

6. Princess Der Ling, *Two Years in the Forbidden City* (New York: Moffat, Yard, 1912; 1911 reprint), 40.

7. Eugene N. Anderson, *The Food of China*, 97–100; and Frederick W. Mote, "Yüan and Ming," in *Food in Chinese Culture: Anthropological and Historical Perspectives*, ed. K. C. Chang (New Haven: Yale University Press, 1977), 196–241.

8. On the issue of Buddhism and fish consumption in Japan, see Simoons, *Eat Not This Flesh*, 288–294.

9. Basil Hall Chamberlain, *Things Japanese, Being Notes on Various Subjects Connected with Japan for the Use of Travellers and Others* (London: John Murray, 1905), 177.

10. Attempting to cite even a few of the relevant sources on the issue of *Buraku* origins and changing social status is foolhardy given how voluminous this literature is. Nevertheless, some of the relevant sources include Hugh Smythe and Yoshimasa Naitoh, "The Eta Caste in Japan," *Phylon* 14, no. 1 (1953): 19–20; John D. Donoghue, "An Eta Community in Japan: The Social Persistence of Outcaste Groups," *American Anthropologist* 59, no. 6 (1957): 1000–1001; Suginohara Juichi, *The Status Discrimination in Japan: Introduction to the Buraku Problem* (Kobe, JP: Hyōgo Institute of Buraku Problems, 1982), chapter 2; Kitaguchi, *An Introduction to the Buraku Issue*, chapter 5; and Ian Neary, "*Burakumin* at the End of History," *Social Research* 70, no. 1 (2003): 269. The Meiji emperor banned use of the designations *eta* and *hinin* in 1871, and the term *heimin* (general people) was substituted. However, use of the designation *Buraku* became more widely used. The payless leather work account is described in Hirai Kiyotaka, *Buraku no Nariwai* [Buraku occupations] (Otsu, JP: Dowa Mondai Kenkyujo, 1993), 127–129.

11. Newman, *Food Culture in China*, 13; Andō Tarō, *Niku Shoku Nitsuite; Meat Eating: An Answer to Some of the Foolish Prejudices Prevailing in Japan Regarding Christianity* (Tokyo: Kyobunkan, 1897); Hirai, *Buraku no Nariwai*, 165, 174, 195–197; Sakurai Atsushi and Kishi Mamoru, eds., *Tojo Bunka* [Slaughterhouse culture] (Tokyo: Sodosha, 2001), 19–34. Krämer, "'Not Befitting Our Divine Country,'" 39.

12. Shimizu Akira, "Meat-Eating in the Kōjimachi District of Edo," in *Japanese Foodways: Past and Present*, ed. Eric C. Rath and Stephanie Assmann (Urbana: University of Illinois Press, 2010), 102–103; see also 99–103.

13. Brett L. Walker, "Commercial Growth and Environmental Change in Early Modern Japan: Hachinohe's Wild Boar Famine of 1749," *Journal of Asian Studies* 60, no. 2

(2001): 342; Susan B. Hanley, *Everyday Things in Premodern Japan: The Hidden Legacy of Material Culture* (Berkeley: University of California Press, 1997), 65–67, 75; Emiko Ohnuki-Tierney, "We Eat Each Other's Food to Nourish Our Body: The Global and the Local as Mutually Constituent Forces," in *Food in Global History*, ed. Raymond Grew (Boulder, CO: Westview, 1999), 252; Ishige, *History and Culture of Japanese Food*, 53–54; Cwiertka, *Modern Japanese Cuisine*, 33; and Fukuhara Yasuo, *Nihon Shokunikushi* [History of Japanese meat] (Tokyo: Shooei-seihan, 1956), 53–54, 112-116.

14. Sakurai and Kishi, *Tojo Bunka*, 20–21; Kamata Satoshi, *Documento Tojo* [Slaughterhouse document] (Tokyo: Iwanami Shoten, 1998), 86; Hirai, *Buraku no Nariwai*, 189; Fukuhara, *Nihon Shokunikushi*, 61, 73–76, 186; Ogura Takekazu, ed., *Agricultural Development in Modern Japan* (Tokyo: Fuji, 1967), 567; Yanagita Kunio, ed. and comp., *Japanese Manners and Customs in the Meiji Era*, trans. Charles S. Terry (Tokyo: Obunsha, 1957), 38; Ishige, *History and Culture of Japanese Food*, 148; Cwiertka, *Modern Japanese Cuisine*, 29; and Penelope Francks, *The Japanese Consumer: An Alternative Economic History of Modern Japan* (Cambridge: Cambridge University Press, 2009), 16.

15. Isabella Bird, *Unbeaten Tracks in Japan* (New York: Putnam's Sons, 1880), 1:53.

16. Ibid., 1:53n1, 1:245–246, 1:280, 2:90.

17. Hayami Yujiro, *Japanese Agriculture under Siege: The Political Economy of Agricultural Policies* (New York: St. Martin's, 1988), 25, 39; and Ogura, *Agricultural Development in Modern Japan*, 566.

18. De Ganon, "The Animal Economy," 135; Ogura, *Agricultural Development in Modern Japan*, 567–580; Penelope Francks with Johanna Boestel and Choo Hyop Kim, *Agriculture and Economic Development in East Asia: From Growth to Protectionism in Japan, Korea and Taiwan* (London: Routledge, 1999), 53, 69; Hayami, *Japanese Agriculture under Siege*, 32; and John Lossing Buck, *Land Utilization in China: A Study of 16,786 Farms in 168 Localities, and 38,256 Farm Families in Twenty-Two Provinces in China, 1929–1933* (New York: Paragon Book Reprint, 1964), 250–251. According to Buck's research in 1929, 14.8 percent of farms had chickens for meat. Only 3 percent raised hogs for meat. Overall, Japanese farms had a smaller percentage of meat animals than China in the same period (18.8 percent versus 25.3 percent). Compared to China, Japan had a slightly higher percentage of farms that included draft animals (81.2 versus 74.7). Horses made up the largest proportion of draft animals (42 percent), and cattle made up the rest (39.2 percent).

19. John M. Maki, *A Yankee in Hokkaido: The Life of William Smith Clark* (Lanham, MD: Lexington, 2002), esp. 183–196.

20. Suginohara, *Status Discrimination in Japan*, 18–19.

21. Fukuhara, *Nihon Shokunikushi*, 79–80, 91; Cwiertka, *Modern Japanese Cuisine*, 24, 29; Yanagita, *Japanese Manners and Customs in the Meiji Era*, 40; de Ganon, "The Animal Economy," 135; Smil and Kobayashi, *Japan's Dietary Transition and Its Impacts*, 49; Richard M. Jaffe, *Neither Monk nor Layman*, 72; and Donald Keene, *Emperor of Japan: Meiji and His World, 1852–1912* (New York: Columbia University Press, 2002), 223.

22. Hirai, *Buraku no Nariwai*, 192–195; and Fukuhara, *Nihon Shokunikushi*, 80. Hirai claims that even during the Taishō Era many Japanese still refused to eat meat. See 207–209.

23. Krämer, "'Not Befitting Our Divine Country,'" 41; Fukuhara, *Nihon Shokunikushi*, 74–75, 99–102, 105; Cwiertka, *Modern Japanese Cuisine*, 18–23; Yanagita, *Japanese Manners and Customs in the Meiji Era*, 40; Ohnuki-Tierney, "We Eat Each Other's Food to Nourish Our Body," 252–253; and Susan B. Hanley, "The Material Culture: Stability in

Transition," in *Japan in Transition: From Tokugawa to Meiji*, ed. Marius B. Jansen and Gilbert Rozman (Princeton: Princeton University Press, 1986), 456.

24. Quoted from a 1966 translation of *Aguranabe*'s introduction in John Pierre Mertz, *Novel Japan: Spaces of Nationhood in Early Meiji Narrative, 1870–88* (Ann Arbor: Center for Japanese Studies, University of Michigan Press), 1. See also Kanagaki Robun, "The Beefeater," in *Modern Japanese Literature: An Anthology*, ed. Donald Keene (New York: Grove, 1956), 31–33; and Fukuhara, *Nihon Shokunikushi*, 102.

25. Mertz, *Novel Japan*, 3–4; de Ganon, "The Animal Economy," 135–136; and Andō, *Niku Shoku Nitsuite.*

26. Cwiertka, *Modern Japanese Cuisine*, 61–65; de Ganon, "The Animal Economy," 157; Ogura, *Agricultural Development in Modern Japan*, 575, 583–584; and Clemen, *American Livestock and Meat Industry*, 283.

27. Fukuhara, *Nihon Shokunikushi*, 81–84.

28. Ibid., 87–89.

29. Ibid., 89–90.

30. Ibid., 112–115.

31. Fukuhara, *Nihon Shokunikushi*, 84–85, 97, 116–119, 185–192; and Yagi Tadashi, "Nihon no shokuniku sangyo ni okeru koyou keitai to no ijtsujou: Buraku sabetsu to shokugyo sabetsu roudou" [Employment relations and discriminative work situations in Japanese meat industries], *Journal of Dowa-Mondai* 17 (1995): 13.

32. Yagi Tadashi, "Shokunikugyo: Shokuniku rōdō no jyugyō jissen ni mukete" [Meat work project report], *Buraku Kaihō Jinken kenkyū hōkokusho* [Buraku liberation and human rights research institute journal], no. 13 (2009): 49–53.

33. Fukuhara, *Nihon Shokunikushi*, 196. Also see John Knight, "Japanese Requiem for British Cows?," *Anthropology Today* 12, no. 6 (1996): 18–20.

34. Fukuhara, *Nihon Shokunikushi*, 98–99, 111, 116; Hirai, *Buraku no Nariwai*, 204; and Francks, *The Japanese Consumer*, 94.

35. John Lossing Buck, *Chinese Farm Economy: A Study of 2866 Farms in Seventeen Localities and Seven Provinces in China* (Chicago: University of Chicago Press, 1930), 378.

36. Jonathan Spence, "Ch'ing," in *Food in Chinese Culture: Anthropological and Historical Perspectives*, ed. K. C. Chang (New Haven: Yale University Press, 1977), 269; Buck, *Chinese Farm Economy*, 356–372; Simoons, *Food in China*, 476; and Vera Y. N. Hsu and Francis L. K. Hsu, "Modern China: North," in *Food in Chinese Culture: Anthropological and Historical Perspectives*, ed. K. C. Chang (New Haven: Yale University Press, 1977), 302.

37. Buck, *Land Utilization in China*, 265. See also 246–256.

38. See Earl B. Shaw, "Swine Industry of China," *Economic Geography* 14, no. 4 (1938): 381–397. For meat consumption estimates in China during the mid-1930s, see Simoons, *Eat Not This Flesh*, 4.

39. Russell H. Chittenden, *The Nutrition of Man* (New York: Stokes, 1907), 223.

## Chapter 3

1. Justus von Liebig, *Animal Chemistry, or Organic Chemistry in Its Application to Physiology and Pathology* (New York: Johnson Reprint, 1964; facsimile of the Cambridge edition of 1842), 48.

2. Ibid., 101. For an in-depth analysis of *Animal Chemistry*, see William H. Brock, *Justus von Liebig: The Chemical Gatekeeper* (Cambridge: Cambridge University Press,

1997), 183–214. On the scientific work preceding Liebig's dealing with protein, as well as Liebig's conclusions in *Animal Chemistry*, see Kenneth J. Carpenter, *Protein and Energy: A Study of Changing Ideas in Nutrition* (Cambridge: Cambridge University Press, 1994), 17–54. On Liebig's influence on agricultural science broadly defined, see Margaret W. Rossiter, *The Emergence of Agriculture Science: Justus Liebig and the Americans, 1840–1880* (New Haven: Yale University Press, 1975), 26–27, 41. Also see Mark R. Finlay, "Early Marketing of the Theory of Nutrition: The Science and Culture of Liebig's Extract of Meat," in *The Science and Culture of Nutrition, 1840–1940*, ed. Harmke Kamminga and Andrew Cunningham (Amsterdam: Rodopi B.V., 1995), 50–51.

3. Brock, *Justus von Liebig*, 222–224; W. A. Shenstone, *Justus von Liebig: His Life and Work (1803–1873)* (New York: Macmillan, 1905), 164–165; and Teuteberg, "The General Relationship between Diet and Industrialization," 91–92.

4. Brock, *Justus von Liebig*, 224–229, 328; Crossley, "Location and Development," 54–65, 71, 75–77, 86–89, 116, 136–137, 153, 183; Graham-Yooll, *The Forgotten Colony*, 227–228; and Hanson, *Argentine Meat and the British Market*, 31–32.

5. Thudichum, *On the Origin, Nature, and Uses of Liebig's Extract*, 7–10, 16–23.

6. Crossley, "Location and Development," 69–111.

7. *Liebig Company's Practical Cookery Book: A Collection of New and Useful Recipes in Every Branch of Cookery* (London: Liebig's Extract of Meat Co., Ltd., 1893), v–vii and inside back cover.

8. See Finlay, "Early Marketing of the Theory of Nutrition," 48–74.

9. Crossley, "Location and Development," 152–153.

10. Ibid., 188–198, 210, 223, 266, 312–320, 328, 332–335, 342–349, 434, 466, 604. Also see Thaddeus Sunseri, "A Political Ecology of Beef in Colonial Tanzania and the Global Periphery, 1864–1961," *Journal of Historical Geography* 39, no. 1 (2013): 32–33. In 1905, both the Fray Bentos and Colón factories were expected to slaughter 20,000 cattle for Oxo and corned beef and 170,000 cattle for extract. Global demand for extract continued to expand during the first decade of the twentieth century. Liebig's produced 1.8 million pounds of it in 1905 and 2.9 million pounds in 1907; the percentage of beef slaughtered for corned beef at Fray Bentos increased from 61 percent in 1915 to 86 percent in 1919; during the 1920s, Liebig's domestic share of its sales jumped from 54 to 68 percent. The Zeballos Cué plant produced corned beef into the 1970s.

11. Carpenter, *Protein and Energy*, 73–74, 89–94; Corinna Treitel, "Food Science/Food Politics: Max Rubner and 'Rational Nutrition' in Fin-de-Siècle Berlin," in *Food and the City in Europe since 1800*, ed. Peter J. Atkins, Peter Lummel, and Derek J. Oddy (Aldershot, UK: Ashgate, 2007), 51–58; Alain Drouard, "Reforming Diet at the End of the Nineteenth Century in Europe," in *Food and the City in Europe since 1800*, ed. Peter J. Atkins, Peter Lummel, and Derek J. Oddy (Aldershot, UK: Ashgate, 2007), 215–218; Charlotte Biltekoff, *Eating Right in America: The Cultural Politics of Food and Health* (Durham, NC: Duke University Press, 2013), 16–17; and Nick Cullather, *The Hungry World: America's Cold War Battle against Poverty in Asia* (Cambridge, MA: Harvard University Press, 2010), 12–15. On the history of the human protein requirement debate, also see Vaclav Smil, *Feeding the World: A Challenge for the Twenty-First Century* (Cambridge: MIT Press, 2000), 228.

12. Naomi Aronson, "Nutrition as a Social Problem: A Case Study of Entrepreneurial Strategy in Science," *Social Problems* 29, no. 5 (June 1982), 474–487; Carpenter, *Protein and Energy*, 102–107; Nick Cullather, "American Pie: The Imperialism of the Calorie," *History Today* 57, no. 2 (February 2007): 34–40; Cullather, *Hungry World*, 11–12;

Harvey Levenstein, *Revolution at the Table: The Transformation of the American Diet* (Berkeley: University of California Press, 2003), 98–108.

13. Wilbur O. Atwater, *Food and Diet*, US Department of Agriculture (Washington, DC: GPO, 1895), 381. See also 357–386.

14. See Levenstein, *Revolution at the Table*, 56, 72–85, 101–102.

15. Atwater, *Food and Diet*, 367.

16. Ibid., 381.

17. Ibid., 553–557, esp. Table D.

18. Carpenter, *Protein and Energy*, 103, 107–112; "Russell Henry Chittenden: February 18, 1856–December 26, 1943: An Appreciation," *Journal of Nutrition* 28, no. 1 (July 1944): 2–6; and L. Margaret Barnett, "The Impact of 'Fletcherism' on the Food Policies of Herbert Hoover during World War I," *Bulletin of the History of Medicine* 66, no. 2 (Summer 1992): 247–258. Interestingly, one of Chittenden's test subjects was Horace Fletcher, the so-called Great Masticator, who during the first decade of the twentieth century promoted the importance of thoroughly chewing—masticating—one's food as a path to better health. In fact, Chittenden's attempts to influence caloric standards during this period and through his later involvement in the US Food Administration during World War I were based in part on his observations of Fletcher. For the current USDA Food Patterns, see www.cnpp.usda.gov/sites/default/files/usda_food_patterns/USDAFood PatternsSummaryTable.pdf. For the USDA Dietary Guidelines, 2015–2020, see health .gov/dietaryguidelines/2015/.

19. Chittenden, *Nutrition of Man*, 127.

20. Ibid., 229.

21. Ibid., 230.

22. Ibid., 226.

23. Ibid., 301. A comparable perspective on the importance of a balanced, economical diet from this era is Whitman H. Jordan, *Principles of Human Nutrition* (New York: Macmillan, 1912), esp. 256–257.

24. James Gregory, *Of Victorians and Vegetarians: The Vegetarian Movement in Nineteenth-Century Britain* (London: Tauris Academic Studies, 2007), esp. 88–97; Adam D. Shprintzen, *The Vegetarian Crusade: The Rise of an American Reform Movement, 1817–1921* (Chapel Hill: University of North Carolina Press, 2013), 10–27; 59–62, 77–78, 91, 94–113; Carpenter, *Protein and Energy*, 77–84; and Cullather, *Hungry World*, 15.

25. Gerald Carson, *Cornflake Crusade* (New York: Rinehart, 1957), 22; Barbara Haber, *From Hardtack to Home Fries: An Uncommon History of American Cooks and Meals* (New York: Free Press, 2002), 69–77; Karen Iacobbo and Michael Iacobbo, *Vegetarian America: A History* (Westport, CT: Praeger, 2004), 111–112; Mary Butler, Frances Thornton, and Garth "Duff" Stoltz, *The Battle Creek Idea: Dr. John Harvey Kellogg and the Battle Creek Sanitarium* (Battle Creek, MI: Heritage, 1994), 6–8; Benjamin Kline Hunnicutt, *Kellogg's Six-Hour Day* (Philadelphia: Temple University Press, 1996), 39–40; Carpenter, *Protein and Energy*, 84–88; and Brian C. Wilson, *Dr. John Harvey Kellogg and the Religion of Biologic Living* (Bloomington: Indiana University Press, 2014), 46–48.

26. John Harvey Kellogg, *Shall We Slay to Eat?* (Battle Creek, MI: Good Health, 1899), 126–127.

27. Ibid., 155. See also 32, 59–90, 114, 126–136, 144–155; and Wilson, *Dr. John Harvey Kellogg and the Religion of Biologic Living*, 47–48, 73. For a complete list of Kellogg's publications, see the online finding aid for the published works of John Harvey Kellogg in

the Bentley Historical Library, University of Michigan Libraries: quod.lib.umich.edu/b/bhlead/umich-bhl-0046?rgn=main;view=text.

28. Irving Fisher, *The Influence of Flesh-Eating on Endurance* (Battle Creek, MI: Modern Medicine, 1908), 1–2, 18–19; William J. Barber, "Irving Fisher of Yale," *American Journal of Economics and Sociology* 64, no. 1 (January 2005): 48–50, 52; Harvey Green, *Fit for America: Health, Fitness, Sport and American Society* (New York: Pantheon, 1986), 300; James Moore, "The Fortunes of Eugenics," in *Medicine Transformed: Health, Disease and Society in Europe, 1800–1930*, ed. Deborah Brunton (Manchester, UK: Manchester University Press), 266–297; and Le, *100 Million Years of Food*, 33.

29. Daniel E. Bender, *American Abyss: Savagery and Civilization in the Age of Industry* (Ithaca, NY: Cornell University Press, 2009), 198–200; Wilson, *Dr. John Harvey Kellogg and the Religion of Biologic Living*, 155–170; Stefan Kühl, *For the Betterment of the Race: The Rise and Fall of the International Movement for Eugenics and Racial Hygiene* (New York: Palgrave Macmillan, 2013), 29–52; Gerald V. O'Brien, *Framing the Moron: The Social Construction of Feeble-Mindedness in the American Eugenic Era* (Manchester, UK: Manchester University Press, 2013), 61.

30. Irving Fisher, "Impending Problems of Eugenics," *Scientific Monthly* 13, no. 3 (September 1921): 220.

31. Kühl, *For the Betterment of the Race*, 52.

32. *Good Health* 54, no. 3 (1919): 130.

33. Woods Hutchinson, *A Handbook of Health*, rev. ed. (Boston: Houghton Mifflin, 1920), 21, 28, 32.

34. John Harvey Kellogg, *The New Dietetics: What to Eat and How: A Guide to Scientific Feeding in Health and Disease* (Battle Creek, MI: Modern Medicine, 1921), 357.

35. Hutchinson, *A Handbook of Health*, 55.

36. Kellogg, *The New Dietetics*, 262. See also 295, 307.

37. Hutchinson, *A Handbook of Health*, 48, 47. See also 56–58.

38. Ibid., 21.

39. Chittenden, *Nutrition of Man*, 163; see also Carpenter, *Protein and Energy*, 117.

40. Woods Hutchinson, *Instinct and Health* (New York: Dodd, Mead, 1908), 21, 31, 33.

41. Ibid., 34.

42. Hutchinson, *A Handbook of Health*, 25.

43. Woods Hutchinson, *Civilization and Health* (Boston: Houghton Mifflin, 1914), 299, 321. See also Hutchinson, "The Importance of Negative Eugenics or the Prevention of Ill-Bornness," *American Journal of Public Health* 3, no. 3 (1913): 238–243; and Moore, "The Fortunes of Eugenics," 272.

44. Kellogg, *The New Dietetics*, 301.

45. Chittenden, *Nutrition of Man*, 222.

46. Patrick Manson, quoted in Shang-Jen Li, "Eating Well in China: Diet and Hygiene in Nineteenth-Century Treaty Ports," in *Health and Hygiene in Chinese East Asia: Policies and Publics in the Long Twentieth Century*, ed. Angela Ki Che Leung and Charlotte Furth (Durham, NC: Duke University Press, 2010), 117.

47. See Douglas M. Haynes, *Imperial Medicine: Patrick Manson and the Conquest of Tropical Disease* (Philadelphia: University of Pennsylvania Press, 2013), 2–4; and Michael Worboys, "Colonial and Imperial Medicine," in *Medicine Transformed: Health, Disease and Society in Europe, 1800–1930*, ed. Deborah Brunton (Manchester, UK: Manchester University Press), 226–229.

48. Kenneth J. Carpenter, *Beriberi, White Rice, and Vitamin B: A Disease, a Cause, and a Cure* (Berkeley: University of California Press, 2000), 2–17; Cwiertka, *Modern Japanese Cuisine*, 64–79; and Ogura, *Agricultural Development in Modern Japan*, 575. Improving the nutritional quality of soldiers' diets was important in Western nations' histories. In United States history, John C. Calhoun advocated dietary improvements for soldiers during his service as secretary of war from 1817 to 1825. See Richard W. Barsness, "John C. Calhoun and the Military Establishment, 1817–1825," *Wisconsin Magazine of History* 50, no. 1 (1966): 50.

49. Cwiertka, *Modern Japanese Cuisine*, 79.

50. A catalog of Liebig's product trading cards can be examined online at www.chromo.be/alfalist.asp.

51. S. Ryan Johansson and Carl Mosk, "Exposure, Resistance and Life Expectancy: Disease and Death during the Economic Development of Japan, 1900–1960," *Population Studies* 41, no. 2 (1987): 207–235; and de Ganon, "The Animal Economy," 138–149.

52. De Ganon, "The Animal Economy," 157–158.

53. R. Iwazumi, "Heavy Odds Oppose Japan's Animal Industry," *Trans-Pacific* 4, no. 5 (May 1921): 68.

54. Robert James McFall, *The World's Meat* (New York: D. Appleton, 1927), 348.

55. Ibid., 330–341, 346–349.

## Chapter 4

1. On heightened government concern about epizootics tied to increased meat consumption, see Karl Appuhn, "Ecologies of Beef: Eighteenth-Century Epizootics and the Environmental History of Early Modern Europe," *Environmental History* 15, no. 2 (2010): 268–287. On bovine tuberculosis, see Waddington, *The Bovine Scourge*, 17–27; Waddington, "'Dead Meat' Dramas: Diseased Meat and the Public's Health," *History Compass* 1 (2003): 1–7; and Waddington, "'Unfit for Human Consumption': Tuberculosis and the Problem of Infected Meat in Late Victorian Britain," *Bulletin of the History of Medicine* 77 (2003): 636–661.

2. Waddington, *The Bovine Scourge*, 11–16, 66; P. J. Atkins, "The Glasgow Case: Meat, Disease and Regulation, 1889–1924," *Agricultural History Review* 52, no. 2 (2004): 167; Deborah Brunton, "Access to Health Care, 1880–1930," in *Medicine Transformed: Health, Disease and Society in Europe, 1800–1930*, ed. Deborah Brunton (Manchester, UK: Manchester University Press), 366; and Uwe Spiekermann, "Dangerous Meat? German-American Quarrels over Pork and Beef, 1870–1900," *Bulletin of the German Historical Institute* 46 (2010), 106.

3. Waddington, *The Bovine Scourge*, 70–91, 148–151; and Atkins, "The Glasgow Case," 164–166, 180–181.

4. Waddington, *The Bovine Scourge*, 135–136, 175–190; and Smith-Howard, *Pure and Modern Milk*, 28–34.

5. Clive A. Spinage, *Cattle Plague: A History* (New York: Kluwer Academic/Plenum, 2003), 5, 103–131, 161–188, 189–216, 241–285; and Dorothee Brantz, "'Risky Business': Disease, Disaster and the Unintended Consequences of Epizootics in Eighteenth- and Nineteenth-Century France and Germany," *Environment and History* 17, no. 1 (2011): 35–51.

6. Brantz, "'Risky Business,'" 44–47.

7. Claire Strom, *Making Catfish Bait out of Government Boys: The Fight against Cattle Ticks and the Transformation of the Yeoman South* (Athens: University of Georgia Press, 2009), 2–5, 13–18, 20–34, 43–73, 76–77, 104, 165–178, 212; Fred Wilbur Powell, *The Bureau of Animal Industry: Its History, Activities and Organization*, Institute for Government Research, Service Monographs of the United States Government, No. 41 (Baltimore: Johns Hopkins Press, 1927), 2–3; *History of Hog Cholera Research in the US Department of Agriculture, 1884–1960*, Agriculture Information Bulletin No. 241, Agricultural Research Service (Washington, DC: January 1962), iii–vi, 2–5, 95; and Laurie Winn Carlson, *Cattle: An Informal Social History* (Chicago: Ivan R. Dee, 2001), 83–84, 96–97, 99–103.

8. *History of Hog Cholera Research*, 5–10, 12–19, 41–49.

9. Powell, *Bureau of Animal Industry*, 10–11; Louis J. Snyder, "The American-German Pork Dispute, 1879–1891," *Journal of Modern History* 17, no. 1 (1945): 17; Gignilliat, "Pigs, Politics, and Protection," 4–7; and Webb, "Agricultural Protection in Wilhelminian Germany," 309–318.

10. Gignilliat, "Pigs, Politics, and Protection," 3; Snyder, "The American-German Pork Dispute," 18, 25; Suellen Hoy and Walter Nugent, "Public Health or Protectionism? The German-American Pork War, 1880–1891," *Bulletin of the History of Medicine* 63, no. 2 (1989): 198–202, 220; Spiekermann, "Dangerous Meat?," 93–97; and James H. Cassedy, "Applied Microscopy and American Pork Diplomacy: Charles Wardell Stiles in Germany, 1898–1899," *Isis* 62, no. 1 (1971): 6–7. On growing British antagonism toward Germany, as exhibited through pork fears, see Keir Waddington, "'We Don't Want Any German Sausages Here! Food, Fear, and the German Nation in Victorian and Edwardian Britain," *Journal of British Studies* 52 (2013): 1017–1042.

11. Gignilliat, "Pigs, Politics, and Protection," 5; Spiekermann, "Dangerous Meat?," 99, 107–108; Snyder, "The American-German Pork Dispute," 25–26; and H. Ray Gamble, "Trichinae," accessed July 7, 2016, www.aphis.usda.gov/vs/trichinae/docs/fact_sheet.htm.

12. Hunt, "Peasants, Grain Tariffs, and Meat Quotas," 314–315; Hoy and Nugent, "Public Health or Protectionism?," 198, 213–222; Gignilliat, "Pigs, Politics, and Protection," 10–12; Snyder, "The American-German Pork Dispute," 20–28; Cassedy, "Applied Microscopy and American Pork Diplomacy," 7–9; Clemen, *American Livestock and Meat Industry*, 285; and Spiekermann, "Dangerous Meat?," 100–108.

13. Doi Rikuo, Ito Akira, Yamazaki Hiroshi, and Morishima Yasuyuki, "Cystic Hydatidosis—Its Occurrence and Prevention in Japan," *Nihon Koshu Eisei Zasshi (Japanese Journal of Public Health)* 50, no. 11 (2003): 1066–1078.

14. Alan Macfarlane, *The Savage Wars of Peace: England, Japan and the Malthusian Trap* (Oxford: Blackwell, 1997), 297–298; Smil and Kobayashi, *Japan's Dietary Transition and Its Impacts*, 50.

15. Chittenden, *Nutrition of Man*, 225–226.

16. Particularly influential in nearly all assessments of the rise of the public slaughterhouse in Europe and elsewhere is Noëlie Vialles, *Animal to Edible*, trans. J. A. Underwood (Cambridge: Cambridge University Press, 1994; originally published in 1987). Vialles argues that the establishment of the first abattoir in 1806 as a facility that separated slaughtering from butchering signaled "an expression of the profound shift in sensibilities" about the making of meat (19). For the purpose of this discussion, the abattoir was central to the public slaughterhouse movement. On the origins of public ownership of meat facilities in France, see Sydney Watts, "The Grande Boucherie," 13–26; and Sydney Watts, *Meat Matters*. On public ownership of meat facilities in France,

particularly the establishment of La Villette, see Horowitz, Pilcher, and Watts, "Meat for the Multitudes," 1065–1075; MacLachlan, "A Bloody Offal Nuisance," 241–242; and Claflin, "La Villette," 27–45. On the rise of public ownership of slaughterhouses in Germany, see Brantz, "Animal Bodies," 193–215. On the history of public slaughterhouse reform in England, see MacLachlan, "A Bloody Offal Nuisance"; Otter, "Civilizing Slaughter," 29–51; and Richard Perren, "Filth and Profit, Disease and Health: Public and Private Impediments to Slaughterhouse Reform in Victorian Britain," in *Meat, Modernity, and the Rise of the Slaughterhouse*, ed. Paula Young Lee (Durham: University of New Hampshire Press, 2008), 127–150. On the development of a public slaughterhouse in Mexico City, see Jeffrey M. Pilcher, "Abattoir or Packinghouse: A Bloody Industrial Dilemma in Mexico City, c. 1890," *Food & History* 3, no. 2 (2005): 119–143. On the brief public slaughterhouse movement in Canada and the establishment of the Toronto Municipal Abattoir, see MacLachlan, *Kill and Chill*, 126–135.

17. Robert Millward, "European Governments and the Infrastructure Industries, c. 1840–1914," *European Review of Economic History* 8, no. 1 (2004): 3–28. For a contemporary laudatory assessment of the British and German experience with municipal ownership of infrastructure industries, see Frederic C. Howe, "Municipal Ownership: The Testimony of Foreign Experience," *Annals of the American Academy of Political and Social Science* 57 (January 1915): 194–208. For another contemporary article advocating municipal ownership of public utilities in the United States, see Richard T. Ely, "Municipal Ownership of Natural Monopolies," *North American Review* 172, no. 532 (1901): 445–455. On the development of various forms of government intervention in Britain's business structure, see, for instance, William A. Robson, *Nationalized Industry and Public Ownership* (London: George Allen & Unwin, 1960).

18. Alain Corbin, *The Foul and the Fragrant: Odor and the French Social Imagination* (Cambridge, MA: Harvard University Press, 1986), 167.

19. Constance Classen, David Howes, and Anthony Synnott, *Aroma: The Cultural History of Smell* (London: Routledge, 1994), 78–79.

20. Ibid., 54–57, 69, 71, 78–79, 170–171. On the increasing number of private slaughterhouses in late nineteenth century Berlin, see Brantz, "Animal Bodies," 204.

21. For comparisons of state involvement in meat production in the early twentieth century, see McFall, *The World's Meat*, 532–543, 567. On the New Zealand and Australian Meat Producers Boards, see John W. Longworth, "The Problem of Meat Marketing: Are Marketing Boards the Answer?," *Australian Quarterly* 44, no. 1 (1972): 52–63; Forrest Capie, "Australian and New Zealand Competition in the British Market, 1920–39," *Australian Economic History Review* 18, no. 1 (1978): 46–63; G. A. Fleming, "Agricultural Support Policies in a Small Open Economy: New Zealand in the 1920s," *Economic History Review* 52, no. 2 (1999): 334–354; Bruce Curtis and James Reveley, "Producers, Processors and Unions: The Meat Producers Board and Labor Relations in the New Zealand Meat Industry, 1952–1971," *Australian Economic History Review* 41, no. 2 (2001): 135–158. For insight into a farmers' cooperative movement in the slaughtering business, see Bronson, "The Saskatchewan Meat Packing Industry", 24–37.

22. Eva Griffith, "Inside and Outside: Animal Activity and the Red Bull Playhouse, St. John Street," in *A Cultural History of Animals in the Age of Enlightenment*, ed. Matthew Senior (Oxford: Berg, 2007), 101–119; MacLachlan, "A Bloody Offal Nuisance," 227–228, 231, 235–237; and Otter, "Civilizing Slaughter," 32–34.

23. MacLachlan, "A Bloody Offal Nuisance," 241–247; Otter, "Civilizing Slaughter," 31; and Grady, "The Cattle and Meat Trades in Leeds," 133–155.

24. MacLachlan, "A Bloody Offal Nuisance," 229, 249–254; and Grady, "The Cattle and Meat Trades in Leeds," 151.

25. Claflin, "La Villette," 31.

26. Ibid., 27–45.

27. Brantz, "Animal Bodies," 201.

28. Ibid., 193–204.

29. Ibid., 206–214. On the context of Berlin's rapid growth and the accompanying environmental concerns, see Marion W. Gray, "Urban Sewage and Green Meadows: Berlin's Expansion to the South, 1870–1920," *Central European History* 47 (2014): 275–306.

30. George Stewart Brown, "Municipal Ownership of Public Utilities," *North American Review* 182, no. 594 (1906): 707; Clemen, *American Livestock and Meat Industry*, 318–329; Louise Carroll Wade, *Chicago's Pride: The Stockyards, Packingtown, and Environs in the Nineteenth Century* (Urbana: University of Illinois Press, 1987), 191–192; and Warren, *Tied to the Great Packing Machine*, 177. Although federal inspection has been the limit of public involvement in American meat production, discussion among industry leaders about the merits of public slaughterhouses did generate some interest during the World War I era. In 1915, members of the American National Live Stock Association discussed several problems and possible solutions to an array of marketing concerns at their annual meeting in Denver, Colorado. One of the recommendations was establishment of public slaughterhouses at public stockyards. This recommendation was likely connected to widespread concerns about packing companies' major financial interests in privately held stockyards around the country. The Packers and Stockyards Act, passed by Congress in 1921, effectively separated packing companies from controlling interest in stockyards. See Clemen, *American Livestock and Meat Industry*, 661–662.

31. Kamata, *Documento Tojo*, 188; Yagi, "Nihon no shokuniku," 18; and Yagi, "Shokunikugyo," 53–57; Doi, Ito, Yamazaki, and Morishima, "Cystic Hydatidosis," 1066–1078; Francks, *The Japanese Consumer*, 94, 228; and Clemen, *American Livestock and Meat Industry*, 256.

## Chapter 5

1. Hagen, *Anglo-Saxon Food and Drink*, 125.

2. Euan Cameron, ed., *Early Modern Europe: An Oxford History* (Oxford: Oxford University Press, 1999), 243; Margaret Murphy, "Feeding Medieval Cities: Some Historical Approaches," 120; Hagen, *Anglo-Saxon Food and Drink*, 120–121, 125, 131–132, 157, 171, 183; Dyer, "Changes in Diet in the Late Middle Ages," 84–86; Algernon Sidney Bicknell, *Hippophagy: The Horse as Food for Man* (London: W. Ridgway, 1868), 5–8; Simoons, *Eat Not This Flesh*, 169–187; Calvin W. Schwabe, *Unmentionable Cuisine* (Charlottesville: University Press of Virginia, 1979), 157; Cathy Reid, "Hippophagy—In Scotland?," *History Scotland Magazine* 14, no. 5 (2014): 22–26; and Daniel W. Gade, "Horsemeat as Human Food in France," *Ecology of Food and Nutrition* 5 (1976): 1–2.

3. Thirsk, *Food in Early Modern England*, xi.

4. Ibid., 143–150, 169–170, 238–241, 244–245, 253–254; Earl B. Shaw, "Swine Industry of China," 393–394; Zhou Dasheng, "An Anthropological Study of Pork-Eating in Asia," in *Changing Chinese Foodways in Asia*, ed. David Y. H. Wu and Tan Chee-beng (Hong Kong: Chinese University Press, 2001), 257–260; and Massimo Montanari, *Medieval Tastes: Food, Cooking, and the Table*, trans. Beth Archer Brombert (New York: Columbia University Press, 2015), 89–106.

5. Michael S. Nassaney, *The Archaeology of the North American Fur Trade* (Gainesville: University of Florida Press, 2015); Peter C. Welsh, *Tanning in the United States to 1850: A Brief History* (Washington, DC: Smithsonian Institution, 1964), 4–14, 15–21, 26, 50; Jared N. Day, "Butchers, Tanners, and Tallow Chandlers: The Geography of Slaughtering in Early Nineteenth-Century New York City," *Food & History* 3, no. 2 (2005): 91; Hanson, *Argentine Meat and the British Market*, 7–8; and Neary, "*Burakumin* at the End of History," 269.

6. Eileen Power, *The Wool Trade in English Medieval History: Being the Ford Lectures* (Oxford: Oxford University Press, 1941), 20–21; Ephraim Lipson, *The History of the Woollen and Worsted Industries* (London: Frank Cass, 1965; originally published in 1921), 3–7, 215–217; Sven Beckert, *Empire of Cotton: A Global History* (New York: Alfred A. Knopf, 2014), 47–49; Peter Lund Simmonds, *Animal Products: Their Preparation, Commercial Uses, and Value* (New York: Scribner, Welford, and Armstrong, 1877), 45–50, 152–165. Hanson's *Argentine Meat and the British Market* notes that, before the 1880s, Argentine sheep had been bred for wool but that by 1900 mutton from the country had found a place in the British market. See 48–50. Lewis Corey points out that until about 1850 most sheep in the United States had been raised for their wool. See Corey, *Meat and Man: A Study of Monopoly, Unionism, and Food Policy* (New York: Viking, 1950), 96.

7. Montanari, *Medieval Tastes*, 89–106; Andrew Hurley, "Busby's Stink Boat and the Regulation of Nuisance Trades, 1865–1918," in *Common Fields: An Environmental History of St. Louis*, ed. Andrew Hurley (St. Louis: Missouri Historical Society Press, 1997), 145–162; K. L. Fry, "Boiling Down in the 1840s: A Grimy Means to a Solvent End," *Labour History* 25 (1973): 1–18; and Nancy Shoemaker, "Whale Meat in American History," *Environmental History* 10, no. 2 (2005): 269–294.

8. Earl B. Shaw, "Swine Industry of China," 381.

9. Ibid., 381–397; David Hollett, *More Precious than Gold: The Story of the Peruvian Guano Trade* (Madison, NJ: Fairleigh Dickinson University Press, 2008); and Simmonds, *Animal Products*, xvii. On the efforts made in East Asia to collect animal (and human) dung for fertilizer, see Franklin Hiram King, *Farmers of Forty Centuries, or Permanent Agriculture in China, Korea and Japan* (Emmaus, PA: Kodale, 1911), 200.

10. Oddy, *From Plain Fare to Fusion Food*, 55; Perren, *Taste, Trade and Technology*, 123–124; and Benjamin Seebohm Rowntree and May Kendall, *How the Labourer Lives: A Study of the Rural Labour Problem*, reprint ed. (New York: Arno, 1975; originally published in 1913), 47, 72, 84, 89, 103, 170, 216. The geographical locations included in Rowntree and Kendall's study were Oxfordshire, Berkshire, Bedfordshire, Essex, North Riding of Yorkshire, East Riding of Yorkshire, and Leicestershire. Derek J. Oddy confirms in large part the basic pattern described here in his analysis of overall British meat consumption in the 1920s: roast for Sunday dinner, cold roast meat on Monday, meat pie on Tuesday, sausages on Wednesday, liver on Thursday, fish on Friday, cold meats, pork pies, potted meat, tongue, or canned corned beef on Saturday. See Oddy, "From Roast Beef to Chicken Nuggets," 236. Fried fish became popular initially in the Lancashire region in the 1870s before catching on throughout the rest of Britain by 1900. Development of the steam trawler in the 1880s that accelerated white-fish trade propelled British consumers' interest in fish and chips. Before World War I, about 25 percent of the white-fish industry was devoted to fish fryers. This proportion would jump to 60 percent by 1939. By the 1920s, 15 percent of potato consumption went to the making of chips. On fish and chips, see John K. Walton, *Fish and Chips and the British Working Class, 1870–1940* (Leicester, UK: Leicester University Press, 1992), 5–8, 25–27, 41–46.

11. Forrest Capie, "Consumer Preference: Meat in England and Wales, 1920–1938," *Bulletin of Economic Research* 28, no. 2 (1976): 92–93.

12. McFall, *The World's Meat*, 427, 436–437.

13. Hanson, *Argentine Meat and the British Market*, 102, 141; Peter H. Smith, *Politics and Beef in Argentina*, 36, 38, 52; McFall, *The World's Meat*, 457–458, 463; Crossley, "Location and Development," 466–467; and Sunseri, "Political Ecology of Beef," 32–33.

14. Henriette Davidis's 1904 *German National Cookery for American Kitchens: Practical Book of the Art of Cooking as Performed in Germany* was reprinted as *Pickled Herring and Pumpkin Pie: A Nineteenth-Century Cookbook for German Immigrants to America*, 3rd American augmented and illustrated edition (Madison, WI: Max Kade Institute for German-American Studies, 2003 reprint; originally published in 1904). Also see Theodore C. Byerly, *Livestock and Livestock Products* (Englewood Cliffs, NJ: Prentice-Hall, 1964), 360.

15. Sam White, "From Globalized Pig Breeds to Capitalist Pigs," 96–103; McIntosh, *American Food Habits*, 82; and Horowitz, *Putting Meat on the American Table*, 44–46.

16. Catherine McNeur, "The 'Swinish Multitude': Controversies over Hogs in Antebellum New York City," *Journal of Urban History* 37, no. 5 (2011): 640.

17. Ibid., 648; and Ross, "Patterns of Diet and Forces of Production," 190–204.

18. Evan Jones, *American Food*, 17–21.

19. Hilliard, *Hog Meat and Hoecake*, 68.

20. Ibid., 55–59, 92–111; Robert William Fogel and Stanley L. Engerman, *Time on the Cross: The Economics of American Negro Slavery* (Boston: Little, Brown, 1974), 109–115; McFall, *The World's Meat*, 111–112, 118; Clemen, *American Livestock and Meat Industry* , 490–498; and Warren, *Tied to the Great Packing Machine*, 143, 187–189.

21. Jennifer Jensen Wallach, "Dethroning the Deceitful Pork Chop: Food Reform at the Tuskegee Institute," in *Dethroning the Deceitful Pork Chop: Rethinking African American Foodways from Slavery to Obama*, ed. Jennifer Jensen Wallach (Fayetteville: University of Arkansas Press, 2015), 168.

22. Ibid., 169; see also 165–179.

23. On Progressive Era promotions of meat eating and superior cultural status in the United States, see Helen Zoe Veit, *Modern Food, Moral Food: Self-Control, Science, and the Rise of Modern American Eating in the Early Twentieth Century* (Chapel Hill: University of North Carolina Press, 2013), 102–106.

24. Kelly J. Sisson Lessens, "Master of Millions: King Corn in American Culture" (PhD diss., University of Michigan, 2011), 158.

25. Ibid., 157–167; see also G. M. Weber and C. L. Alsberg, *The American Vegetable-Shortening Industry: Its Origin and Development* (Stanford, CA: Food Research Institute, 1934) 8–105; John Whittlesey Bodman, "Chemical Control Work in the N. K. Fairbanks [*sic*] Factories," *Professional Degree Theses*, Paper 41 (1917), 1–18, accessed on July 8, 2015, scholarsmine.mst.edu/professional_theses/41; Rudolf Alexander Clemen, *By-Products in the Packing Industry* (Chicago: University of Chicago Press, 1927), 101–104, 119, 158–159; Mitchell Okun, *Fair Play in the Marketplace: The First Battle for Pure Food and Drugs* (DeKalb: Northern Illinois University Press, 1986), 270–271; William Cronon, *Nature's Metropolis: Chicago and the Great West* (New York: Norton, 1991), 252–253; USDA, Daniel D. Harlan, *Lard Marketing as Affected by Commercial Processing Methods*, Agriculture Information Bulletin No. 53, Production and Marketing Administration (Washington, DC, June 1951); Theodore J. Weiss, *Food Oils and Their Uses* (Westport, CT: AVI, 1970), 126–127.

26. *Hogs and Hog Products: A Survey of United States and Foreign Production and International Trade in Hogs, Pork, and Lard, with Special Reference to Factors Essential to Tariff Consideration*, Report No. 143, Second Series (Washington, DC: United States Trade Commission, 1941), 74–88.

27. Clemen, *By-Products in the Packing Industry*, 126, 127.

28. USDA, W. D. Bigelow, *Pure-Food Laws of European Countries affecting American Exports*, Bulletin No. 61, Division of Chemistry (Washington, DC: GPO, 1901).

29. *Hogs and Hog Products*, 102; and J. L. Anderson, "Lard to Lean: Making the Meat-Type Hog in Post–World War II America," in *Food Chains: From Farmyard to Shopping Cart*, ed. Warren Belasco and Roger Horowitz (Philadelphia: University of Pennsylvania Press, 2009), 29–46.

30. Clemen, *American Livestock and Meat Industry*, 462–478; Warren, *Tied to the Great Packing Machine*, 151–154; and John Ryan, "Statistics of Tins and Cans," *Journal of the Royal Statistical Society*, Series A (General) 115, no. 4 (1952): 487, 492.

31. Masuoka Jitsuichi, "Changing Food Habits of the Japanese in Hawaii," *American Sociological Review* 10, no. 6 (1945): 761; Cwiertka, *Modern Japanese Cuisine*, 33; Ishige, *History and Culture of Japanese Food*, 153; Francks, *The Japanese Consumer*, 94, 125, 228; McFall, *The World's Meat*, 318, 346–349; and Nakase Shinzo, *Japan's Livestock Trade* (Tokyo: Japan International Agricultural Council, March 1989), 3–4.

32. Bicknell, *Hippophagy*, 10–24, 38; Gade, "Horsemeat as Human Food in France," 2–4; Reid, "Hippophagy," 26; Simoons, *Eat Not This Flesh*, 190; Warren, *Tied to the Great Packing Machine*, 134; and Smil, *Should We Eat Meat?*, 66.

33. Clemen, *By-Products in the Packing Industry*, 3–9, 12–76; Clemen, *American Livestock and Meat Industry*, 347, 355–356; Clemen, *Is Leather Reviving? A Study of the Present Conditions in the Hide and Leather Industry* (Chicago: Illinois Merchants Trust, August 1924), 3–14. Also see Corey, *Meat and Man*, 44; Cronon, *Nature's Metropolis*, 250–251; and Byerly, *Livestock and Livestock Products*, 363.

34. Corey, *Meat and Man*, 44; and Clemen, *American Livestock and Meat Industry*, 347–356.

35. Simmonds, *Animal Products*, 147–152; Clemen, *By-Products in the Packing Industry*, 117, 145–148, 173, 187–199; Clemen, *American Livestock and Meat Industry*, 358; Corey, *Meat and Man*, 268–269; Whelpley, "American Control of England's Food Supply," 802; Fry, "Boiling Down in the 1840s," 1–18; Shirley Fisher, "The Pastoral Interest and Sydney's Public Health," *Historical Studies* 20, no. 78 (1982): 73–89; Greenhill, "Shipping and the Refrigerated Meat Trade," 68; Sabato, *Agrarian Capitalism and the World Market*, 25, 31; Hanson, *Argentine Meat and the British Market*, 7–9; L. H. Bailey, ed., *Cyclopedia of Farm Animals* (New York: Macmillan, 1922), 259; Walsh, *The Rise of the Midwestern Meat Packing Industry*, 32–33; and Wade, *Chicago's Pride*, 9–10.

36. Clemen, *American Livestock and Meat Industry*, 360–361, 369–378; Clemen, *By-Products in the Packing Industry*, 201–221; Corey, *Meat and Man*, 180; and Naomi Pfeffer, "How Abattoir 'Biotrash' Connected the Social Worlds of the University Laboratory and the Disassembly Line," in *Meat, Medicine and Human Health in the Twentieth Century*, ed. David Cantor, Christian Bonah, and Matthias Dörries (London: Pickering and Chatto, 2010), 65–67.

37. Nils Kessel, "Beyond Innovation: Pharmaceutical Marketing, Market Structure and the Importance of the 'Old' in West Germany, 1950–1970," in *The Development of Scientific Marketing in the Twentieth Century: Research for Sales in the Pharmaceutical Industry*, ed. Jean-Paul Gaudillière and Ulrike Thoms (London: Pickering and Chatto, 2015), 17.

38. Tricia Close-Koenig and Ulrike Thoms, "A Balancing Act: Antidiabetic Products and Diabetes Markets in Germany and France," in *The Development of Scientific Marketing in the Twentieth Century: Research for Sales in the Pharmaceutical Industry*, ed. Jean-Paul Gaudillière and Ulrike Thoms (London: Pickering and Chatto, 2015), 129.

39. Pfeffer, "How Abattoir 'Biotrash' Connected the Social Worlds," 67–73.

40. Clemen, *American Livestock and Meat Industry*, 369–370; and Clemen, *By-Products in the Packing Industry*, 207–221.

41. Corey, *Meat and Man*, 180; and Pfeffer, "How Abattoir 'Biotrash' Connected the Social Worlds," 74–76.

42. Katherine C. Grier, "Provisioning Man's Best Friend: The Early Years of the American Pet Food Industry, 1870–1942," in *Food Chains: From Farmyard to Shopping Cart*, ed. Warren Belasco and Roger Horowitz (Philadelphia: University of Pennsylvania Press, 2009), 127–131; and Grier, *Pets in America: A History* (Chapel Hill: University of North Carolina Press, 2006), 281–285.

43. Grier, "Provisioning Man's Best Friend," 131–141; Grier, *Pets in America*, 285–287; Clemen, *By-Products in the Packing Industry*, 321–328; and Warren, *Tied to the Great Packing Machine*, 198.

44. Clemen, *By-Products in the Packing Industry*, 334–347; William Boyd, "Making Meat: Science, Technology, and American Poultry Production," *Technology and Culture* 42, no. 4 (October 2001): 631–664.

45. Warren, *Tied to the Great Packing Machine*, 196–200; A. M. Pearson and T. R. Dutson, eds., *Edible Meat By-Products: Advances in Meat Research*, vol. 5 (London: Elsevier Applied Science, 1988), 409–425; and Colin G. Brown, John W. Longworth, Scott A. Waldron, *Regionalisation and Integration in China: Lessons from the Transformation of the Beef Industry* (Aldershot, UK: Ashgate, 2002), 44. Timothy Pachirat, *Every Twelve Seconds: Industrialized Slaughter and the Politics of Sight* (New Haven: Yale University Press, 2011), appendix B, provides a detailed list of current beef by-products.

## Chapter 6

1. The FAO, organized in 1943 and made an agency of the United Nations in 1945, provides the most systematic array of statistics related to global food production and consumption. For meat, FAO statistics begin in 1961. However, as Vaclav Smil explains at length in *Should We Eat Meat?*, 102–103, the food balance data included in the FAO statistics do not indicate actual *consumption* but how much meat is *supplied* (emphasis added). That is, the statistics include amounts of meat that are wasted in the supply chain. Thus, the FAO data cited are somewhat inflated proxies for actual consumption. But given that they are relatively uniform and exist for all countries, the term consumption is used throughout the chapter instead of food supplied. All statistics, unless otherwise noted, provided in this chapter come from the online FAOSTAT database: faostat3.fao.org/download/FB/CL/E.

2. See, especially, Friedmann, "The Political Economy of Food: The Rise and Fall," S248–S286; Ross B. Talbot, "Food in the American Political Economy," *Proceedings of the Academy of Political Science* 34, no. 3 (1982): 1–11; Friedmann and McMichael, "Agriculture and the State System," 93–117; Philip McMichael, "Tensions between National and International Control of the World Food Order," 343–365; Harriet Friedmann, "Distance and Durability: Shaky Foundations of the World Food Economy," *Third World Quarterly* 13, no. 2 (1992): 371–383; Harriet Friedmann, "The Political Economy of Food: A

Global Crisis," *New Left Review* 197 (1993): 29–57; and Hugh Campbell and Jane Dixon, "Introduction to the Special Symposium: Reflecting on Twenty Years of the Food Regimes Approach in Agri-Food Studies," *Agriculture and Human Values* 36, no. 4 (2009): 261–265. On GATT, see Susan Ariel Aaronson, *Trade and the American Dream: A Social History of Postwar Trade Policy* (Lexington: University Press of Kentucky, 1996); Thomas W. Zeiler, *Free Trade, Free World: The Advent of GATT* (Chapel Hill: University of North Carolina Press, 1999); Douglas A. Irwin, Petros C. Mavroidis, Alan O. Sykes, *The Genesis of the GATT* (Cambridge: Cambridge University Press, 2008); and Soo Yeon Kim, *Power and the Governance of Global Trade: From the GATT to the WTO* (Ithaca: Cornell University Press, 2010). Statistics are from Aaronson, *Trade and the American Dream*, 16; and Kim, *Power and the Governance of Global Trade*, 90. On developing countries' trade shifts, see H. B. Malmgren and D. L. Schlechty, "Technology and Neo-Mercantilism in International Agricultural Trade," *American Journal of Agricultural Economics* 51, no. 5 (1969), 1328, 1330.

3. Quentin M. West, "World Trade Prospects for U.S. Agriculture," *American Journal of Agricultural Economics* 54, no. 5 (1972): 828–829; Raymond F. Hopkins and Donald J. Puchala, "Perspectives on the International Relations of Food," *International Organizations* 32, no. 3 (1978): 586–588; Dale E. Hathaway, "Changing Patterns of World Trade," *American Journal of Agricultural Economics* 61, no. 5 (1979): 1016–1018; Harry D. Fornari, "U.S. Grain Exports: A Bicentennial Overview," *Agricultural History* 50, no. 1 (1976): 137–150; and Alexander H. Sarris and Andrew Schmitz, "Toward a U.S. Agricultural Export Policy for the 1980s," *American Journal of Agricultural Economics* 63, no. 5 (1981): 833.

4. C. M. Purves, "Major Shifts in World Agriculture," *Journal of Farm Economics* 27, no. 2 (1945): 247; Perrin, *Taste, Trade and Technology*, 157–176; James Murton, "John Bull and Sons: The Empire Marketing Board and the Creation of a British Imperial Food System," in *Edible Histories, Cultural Politics: Towards a Canadian Food History*, ed. Franca Iacovetta, Valerie J. Korinek, and Marlene Epp (Toronto: University of Toronto Press, 2000), 225–248; Curtis and Reveley, "Producers, Processors and Unions," 139; *Australia's Livestock and Meat Industry and the U.S. Producer*, FAS-M-63, US Department of Agriculture, Foreign Agricultural Service, October 1959, 14–23; *New Zealand's Livestock and Meat Industry and the U.S. Producer*, FAS-M-67, US Department of Agriculture, Foreign Agricultural Service, July 1960, 14–19; Friedmann, "The Political Economy of Food: The Rise and Fall," S261; and Paul Conkin, *A Revolution Down on the Farm: The Transformation of American Agriculture since 1929* (Lexington: University Press of Kentucky, 2008), 142.

5. Christina L. Davis, *Food Fights over Free Trade: How International Institutions Promote Agricultural Trade Liberalization* (Princeton: Princeton University Press, 2003), 10.

6. Ibid., 227–251; Andreas Staab, *The European Union Explained: Institutions, Actors, Global Impact*, 3rd ed. (Bloomington: Indiana University Press, 2013), 118–124; Malmgren and Schlechty, "Technology and Neo-Mercantilism in International Agricultural Trade," 1328; Philip McMichael, "Global Development and the Corporate Food Regime," in *New Directions in the Sociology of Global Development*, ed. Frederick H. Buttel and Philip McMichael (Amsterdam: Elsevier, 2005), 277; and Perren, *Taste, Trade and Technology*, 175–179. On the animal products debates in the formation of GATT, see Aaronson, *Trade and the American Dream*, 81–82; Zeiler, *Free Trade, Free World*, 95–103, 114–123; Kim, *Power and the Governance of Global Trade*, 53; and Perren, *Taste, Trade*

*and Technology*, 173. On GATT and agricultural products trade, see Davis, *Food Fights over Free Trade*. For discussion of the politics of Japan's entry into GATT, see Michiko Ikeda, *Japan in Trade Isolation, 1926–37 and 1948–85* (Tokyo: International House of Japan, 2008).

7. Hopkins and Puchala, "Perspectives on the International Relations of Food," 613; and William D. Heffernan and Douglas H. Constance, "Transnational Corporations and the Globalization of the Food System," in *From Columbus to ConAgra: The Globalization of Agriculture and Food*, ed. Alessandro Bonanno, Lawrence Busch, William Friedland, Lourdes Gouveia, and Enzo Mingione (Lawrence: University Press of Kansas, 1994), 29–51.

8. *Trends in U.S. Imports of Manufactured Beef*, FAS-M-73, US Department of Agriculture, Foreign Agricultural Service, November 1959, 2–8; Billie R. DeWalt, "Mexico's Second Green Revolution: Food for Feed," *Mexican Studies/Estudios Mexicanos* 1, no. 1 (1985): 29–60; Ross, "Patterns of Diet and Forces of Production," 204–215; and Chad Broughton, *Boom, Bust, Exodus: The Rust Belt, the Maquilas, and a Tale of Two Cities* (Oxford: Oxford University Press, 2015), 115, 240–242.

9. Deborah Gewertz and Frederick Errington, *Cheap Meat: Flap Food Nations in the Pacific Islands* (Berkeley: University of California Press, 2010), 10, 31–35, 73–74, 105, 115; Miguel Teubal, "Internationalization of Capital and Agroindustrial Complexes: Their Impact on Latin American Agriculture," *Latin American Perspectives* 14, no. 3 (1987): 316–364; Luis Llambi, "Opening Economies and Closing Markets: Latin American Agriculture's Difficult Search for a Place in the Emerging Global Order," in *From Columbus to ConAgra: The Globalization of Agriculture and Food*, ed. Alessandro Bonanno, Lawrence Busch, William Friedland, Lourdes Gouveia, and Enzo Mingione (Lawrence: University Press of Kansas, 1994), 184–209.

10. Philip McMichael, "Global Development and the Corporate Food Regime," 281.

11. David Garst, "What Farmers Want," *Agricultural History* 64, no. 2 (1990): 319–327; Philip McMichael, "Tensions between National and International Control of the World Food Order," 343–365; W. N. Pritchard, "The Emerging Contours of the Third Food Regime: Evidence from Australian Dairy and Wheat Sectors," *Economic Geography* 74, no. 1 (1998): 64–74; Walter P. Falcon and Rosamond L. Naylor, "The Maize Transition in Asia: Unlocking the Controversy," *American Journal of Agricultural Economics* 80, no. 5 (1998): 960–968; Harriet Friedmann, "From Colonialism to Green Capitalism: Social Movements and Emergence of Food Regimes," in *New Directions in the Sociology of Global Development*, ed. Frederick H. Buttel and Philip McMichael (Amsterdam: Elsevier, 2005), 227–264.

12. Philip McMichael, "Global Development and the Corporate Food Regime," 285.

13. Dave M. Juday, "Intensification of Agriculture and Free Trade," in *Livestock, Ethics and Quality of Life*, ed. John Hodges and In K. Han (Wallingford, UK: CABI, 2000), 157–174; and Collingham, *The Taste of War*, 3.

14. Schwabe, *Unmentionable Cuisine*, 215–216; Horowitz, *Putting Meat on the American Table*, 104–110; Page Smith and Charles Daniel, *The Chicken Book* (San Francisco: North Point, 1982), 38–39; and Donald D. Stull and Michael J. Broadway, *Slaughterhouse Blues: The Meat and Poultry Industry in North America* (Belmont, CA: Wadsworth/Thomson Learning, 2004), 37–38.

15. Smith and Daniel, *The Chicken Book*, 232–263; Warren, *Tied to the Great Packing Machine*, 222; McFall, *The World's Meat*, 191, 252–253, 512; and Clemen, *By-Products in the Packing Industry*, 334–347.

16. Boyd, "Making Meat," 635.

17. Ibid., 631–664.

18. Ibid., 636.

19. William Boyd and Michael Watts, "Agro-Industrial Just-in-Time: The Chicken and Postwar American Capitalism," in *Globalising Food: Agrarian Questions and Global Restructuring* (London: Routledge, 1997), 192–225; and Douglas H. Constance, "The Southern Model of Broiler Production and Its Global Implications," *Culture & Agriculture* 30, no. 1–2 (2008): 17–31.

20. Stull and Broadway, *Slaughterhouse Blues*, 39–51; Horowitz, *Putting Meat on the American Table*, 106–128; Heffernan and Constance, "Transnational Corporations and the Globalization of the Food System," 32–40; and Steve Striffler, *Chicken: The Dangerous Transformation of America's Favorite Food* (New Haven: Yale University Press, 2005), 15–70. On John Tyson and his company, see Christopher Leonard, *The Meat Racket: The Secret Takeover of America's Food Business* (New York: Simon and Schuster, 2014).

Before World War II, turkey consumption was even more unusual than eating chicken. Americans ate turkey on a seasonal basis, generally only between Thanksgiving and New Year's. Turkey-processing facilities were typically open only from late September through December. See Hope Nisly, "A Mennonite Woman in 'Thanksgiving Town': Edith Swartzendruber Nisly's Work Experience, 1935–1941," *Labor's Heritage* 3 (1991): 26–48.

21. *Meat & Poultry 2006 Facts* (Washington, DC: American Meat Institute, 2006), 36–37; *Watt PoultryUSA Top Poultry Companies*, accessed August 21, 2015, www.wattagnet.com/articles/21973-watt-poultryusa-top-poultry-companies-mid-size-broiler-producers-lead-industry-in-production-increases?v=preview; and Top Ten Turkey Producing States and Processors, accessed August 21, 2015, www.rockthecapital.com/11/22/top-ten-turkey-producing-states-processors/.

22. Andrew Godley and Bridget Williams, "Democratizing Luxury and the Contentious 'Invention of the Technological Chicken' in Britain," *Business History Review* 83, no. 2 (2009): 267–290; Godley and Williams, "The Chicken, the Factory Farm, and the Supermarket: The Emergence of the Modern Poultry Industry in Britain," in *Food Chains: From Farmyard to Shopping Cart*, ed. Warren Belasco and Roger Horowitz (Philadelphia: University of Pennsylvania Press, 2009), 47–61; and Perren, *Taste, Trade and Technology*, 205–207, 211.

23. Jane Dixon, *The Changing Chicken: Chooks, Cooks and Culinary Culture* (Sydney: University of New South Wales Press, 2002), 3; see also 83–97.

24. Rachel A. Ankeny, "The Moral Economy of Red Meat in Australia," in *Food and Morality*, ed. Susan R. Friedland (Devon, UK: Prospect, 2008), 20–28.

25. Gary Vocke, "Investments to Transfer Poultry Production to Developing Countries," *American Journal of Agricultural Economics* 73, no. 3 (1991): 953; "Thailand's Poultry Industry," n.d., accessed August 24, 2015, www.fao.org/docrep/004/ac797e/ac797e06.htm; and John Giles, "Overview of the World Poultry Sector and the Role of Thailand," GlobalMeatNews.com, January 27, 2012, accessed August 24, 2015, www.globalmeatnews.com/content/view/print/607459.

26. Ancel Keys and Margaret Keys, *Eat Well and Stay Well*, rev. ed. (Garden City, NY: Doubleday, 1963), 154.

27. US Senate, Select Committee on Nutrition and Human Needs, *Dietary Goals for the United States*, 2nd ed. (Washington, DC: GPO, December 1977), 4.

28. Striffler, *Chicken*, 15, 27–30; Horowitz, *Putting Meat on the American Table*, 118, 127–128; Dixon, *Changing Chicken*, 71–79, 150–151; Perren, *Taste, Trade and Technology*, 188–193, 205–208; Schwabe, *Unmentionable Cuisine*, 215–218; Simoons, *Eat Not This Flesh*, 144–161; and Mark Schatzker, *The Dorito Effect: The Surprising New Truth about Food and Flavor* (New York: Simon and Schuster, 2015), 21–22, 26–29, 33–40.

29. Andrew F. Smith, *Fast Food: The Good, the Bad and the Hungry* (London: Reaktion Books, 2016), 17; Eric Schlosser, *Fast Food Nation: The Dark Side of the All-American Meal* (Boston: Houghton Mifflin, 2001), 197–199; Warren, *Tied to the Great Packing Machine*, 156–159; *Meat & Poultry 2006 Facts*, 51; Roberto A. Ferdman, "The Only Meat Everyone Can't Stop Eating," *Washington Post*, March 15, 2016, accessed June 28, 2016, www.washingtonpost.com/news/wonk/wp/2016/03/15/the-incredible-thing-about-hamburgers/.

30. Pablo Pacheco and René Poccard-Chapuis, "Cattle Ranching Development in the Brazilian Amazon: Looking at Long-Term Trends to Explore the Transition towards Sustainable Beef Cattle Production," in *Political Ecologies of Meat*, ed. Jody Emel and Harvey Neo (London: Routledge, 2015), 50; Smil, *Should We Eat Meat?*, 119; Roberts, *The End of Food*, 140; "Big Meat Is Growing in the South," *Seedling*, October 2010, 4–12, accessed August 24, 2015, www.grain.org/es/article/entries/4044-big-meat-is-growing-in-the-south; and Weis, *The Ecological Hoofprint*, 86–87.

31. Yagi, "Shokunikugyo," 65–66; Fukuhara, *Nihon Shokunikushi*, 145-146; 202–203; Jin Pusan, "To Feed a Country at War: China's Supply and Consumption of Grain during the War of Resistance," in *China in the Anti-Japanese War, 1937–1945*, ed. David P. Barrett and Larry N. Shyu (New York: Peter Lang, 2001), 157–169; Dale E. Hathaway, Hendrik S. Houthakker, and John A. Schnittker, "Food Prices and Inflation," *Brookings Papers on Economic Activity* 1974, no. 1 (1974): 91; and Ogura, *Agricultural Development in Modern Japan*, 586.

32. Owen Griffiths, "Need, Greed, and Protest in Japan's Black Market, 1938–1949," *Journal of Social History* 35, no. 4 (Summer 2002): 828–834; John W. Dower, *Embracing Defeat: Japan in the Wake of World War II* (New York: Norton, 1999), 91; and Fukuhara, *Nihon Shokunikushi*, 149–160.

33. Fukuhara, *Nihon Shokunikushi*, 162–169.

34. Hane Mikiso, *Modern Japan: A Historical Survey*, 2nd ed. (Boulder, CO: Westview, 1992), 347–348; Dower, *Embracing Defeat*, 525–546; "Spam-a-Lot," *Asianweek* 2, no. 47 (2006): 14; R. P. Dore, "Beyond the Land Reform: Japan's Agricultural Prospect," *Pacific Affairs* 36, no. 3 (Autumn 1963): 266; Steven K. Vogel, "When Interests Are Not Preferences: The Cautionary Tale of Japanese Consumers," *Comparative Politics* 31, no. 2 (January 1999): 189. On post–World War II goat and rabbit meat production, see *Meat Statistics in Japan*, Livestock Bureau, Ministry of Agriculture and Forestry, June 1, 1977. *The Market in Japan for U.S. Livestock and Meat Products*, US Department of Agriculture, Foreign Agricultural Service, FAS M-78, February 1960, 2–3; Francks, Boestel, and Kim, *Agriculture and Economic Development in East Asia*, 54; and Nakase, *Japan's Livestock Trade*, 5.

35. *Conference on the Advancement of Productivity in Agriculture, Forestry, and Water Industry* (Tokyo: Noorin suisan gyoo seisansei koojyoo kaigi, 1959), 47–58; *The Market in Japan for U.S. Livestock and Meat Products*, 5; *Historical Statistics of Japan*, Statistics Bureau, Ministry of Internal Affairs and Communication, Table 20-1-a: Annual

Average of Monthly Consumption Expenditures per Household (All Households)—All Cities (1946–1962), www.stat.go.jp/english/data/chouki/20.htm; *Historical Statistics of the United States*, Millennial Edition Online, Consumption Expenditures, by Type: 1929–1999, Series Cd153–263, hsus.cambridge.org/HSUSWeb/table/jumpto.do?id=Cd153–263; US Department of Agriculture, *Food Consumption of Households in the United States*, US Agricultural Research Service, Household Economics Research Branch (Washington, DC, December 1956); *The Market in Japan for U.S. Livestock and Meat Products*, 5–15; *New Zealand's Livestock and Meat Industry and the U.S. Producer*, 17–18; and *Australia's Livestock and Meat Industry and the U.S. Producer*, 20, 22.

36. R. L. Sheppard and N. J. Beun, "Japanese Food Policy and Self Sufficiency: An Analysis with Reference to Meat," Discussion Paper No. 47, March 1980, i.

37. Nakase, *Japan's Livestock Trade*, 8.

38. Hayami Yujiro, "Institutional Aspects of Agricultural Development," in *The Agricultural Development of Japan: A Century's Perspective*, ed. Hayami Yujiro and Yamada Saburo (Tokyo: University of Tokyo Press, 1991), 89; Hayami, *Japanese Agriculture under Siege*, 47; James R. Simpson, Yoshida Tadashi, Miyazaki Akira, and Kada Ryohei, *Technological Change in Japan's Beef Industry* (Boulder, CO: Westview, 1985), 23; Thomas I. Wahl, Dermot J. Hayes, and Gary W. Williams, "Dynamic Adjustment in the Japanese Livestock Industry under Beef Import Liberalization," *American Journal of Agricultural Economics* 73, no. 1 (February 1991): 119; and Perren, *Taste, Trade and Technology*, 182.

39. David E. Lindstrom, "Japan Needs U.S. Surplus Food," *Journal of Farm Economics* 371 (February 1955): 125–127; Nakamura Hiroshi, "The Japanese Soybean Market," *Illinois Agricultural Economics* 1, no. 2 (July 1961): 8–9; Friedmann, "The Political Economy of Food: The Rise and Fall," S248–S286; Friedmann, "Distance and Durability: Shaky Foundations of the World Food Economy," 371–383; Nakamura, "The Japanese Soybean Market," 7–12; and Fornari, "U.S. Grain Exports," 148–150.

40. Honma Masayoshi and Hayami Yujiro, "Causes of Growth in Agricultural Protection," in *The Agricultural Development of Japan: A Century's Perspective*, ed. Hayami Yujiro and Yamada Saburo (Tokyo: University of Tokyo Press, 1991), 221; Nakase, *Japan's Livestock Trade*, 4–6; Hathaway, Houthakker, and Schnittker, "Food Prices and Inflation," 90–93; William W. Cochrane, "Food, Agriculture, and Rural Welfare: Domestic Policies in an Uncertain World," *American Journal of Agricultural Economics* 56, no. 5 (December 1974): 991; Gene Gregory, "Japanese Economic Growth: The Human Equation," *Asian Survey* 15, no. 10 (October 1975): 859–860; George Kent, "Dominance in Fishing," *Journal of Peace Research* 13, no. 1 (1976): 39–44; I. M. Destler, "United States Food Policy 1972–1976: Reconciling Domestic and International Objectives," *International Organization* 32, no. 3 (Summer 1978): 628; and Robert L. Paarlberg, "Shifting and Sharing Adjustment Burdens: The Role of the Industrial Food Importing Nations," *International Organization* 32, no. 3 (Summer 1978): 657–659, 664–669.

41. Sheppard and Beun, "Japanese Food Policy and Self-Sufficiency," 6; Perren, *Taste, Trade and Technology*, 183; Cheryl Christensen, "World Hunger: A Structural Approach," *International Organization* 32, no. 3 (Summer 1978): 745–774; Henry R. Nau, "The Diplomacy of World Food: Goals, Capabilities, Issues and Arenas," *International Organization* 32, no. 3 (Summer 1978): 775–809; George J. Viksnins, "U.S.-Japanese Trade: Perceptions and Reality," *Asian Survey* 19, no. 3 (March 1979): 205–229; Hayami Yujiro "Trade Benefits to All: A Design of the Beef Import Liberalization in Japan," *American Journal of Agricultural Economics* 61, no. 2 (May 1979): 342–347; James P. Houck,

"Agricultural Trade: Protectionism, Policy, and the Tokyo/Geneva Negotiating Round," *American Journal of Agricultural Economics* 61, no. 5 (December 1979): 860–873; Ogata Sadako, "Some Japanese Views on United States–Japan Relations in the 1980s," *Asian Survey* 20, no. 7 (July 1980): 694–706; Frances M. Ufkes, "Trade Liberalization, Agro-Food Politics and the Globalization of Agriculture," *Political Geography* 12, no. 3 (May 1993); 215–231; Nakase, *Japan's Livestock Trade*, 20–21; Sato Kazuo, "Japan's Resource Imports," *Annals of the American Academy of Political and Social Science* 513 (January 1991): 76–89; Yoshimatsu Hidetaka, "Japan's Keidanren and Political Influence on Market Liberalization," *Asian Survey* 38, no. 3 (March 1998): 328–345; Wahl, Hayes, and Williams, "Dynamic Adjustment in the Japanese Livestock Industry," 118–132; Nakase, *Japan's Livestock Trade*, 6, 17–18, 24–30; and David Vogel, "Consumer Protection and Protectionism in Japan," *Journal of Japanese Studies* 18, no. 1 (Winter 1992): 119–154.

42. Francks, *The Japanese Consumer*, 165; Morieda Takashi, "The Unlikely Love Affair with Curry and Rice," *Japan Quarterly* 47 (April-June 2000): 66–73; Ufkes, "Trade Liberalization, Agro-Food Politics and the Globalization of Agriculture," 226–228; Heffernan and Constance, "Transnational Corporations and the Globalization of the Food System," 29–51; Perren, *Taste, Trade and Technology*, 183; John Fraser Hart and Chris Mayda, "Pork Palaces on the Panhandle," *Geographical Review* 87, no. 3 (July 1997): 396–400; and Michael F. Beer, *Report of Meat Industry Tour to United States, Japan and Republic of Korea*: A NSW [New South Wales] Agriculture Study Tour Report (Tamworth, AU: NSW Agriculture, 1999).

43. Kaneko Nubuo, *Tanoshii Yuushoku* (*Fun Dinner*) (Tokyo: Jitsugyo, 1990). On the persistence of "wet" preparations of meat in Japanese cuisine, see, for example, Simpson, Yoshida, Miyazaki, and Kada, *Technological Change in Japan's Beef Industry*, 26.

44. Ikai Takeshi, "Global Hamburgers in Japan, 1971–2004: 'Re-Invention of America in Japan' in the Burger Companies" (master's thesis, Doshisha University, 2005); Cwiertka, *Modern Japanese Cuisine*, 164–166; Perren, *Taste, Trade and Technology*, 212; Emiko Ohnuki-Tierney, *Rice as Self: Japanese Identities through Time* (Princeton: Princeton University Press, 1993); Ohnuki-Tierney, "McDonald's in Japan: Changing Manners and Etiquette," in *Golden Arches East: McDonald's in East Asia*, ed. James L. Watson (Stanford: Stanford University Press, 1997), 166–169; John W. Traphagan and L. Keith Brown, "Fast Food and Intergenerational Commensality in Japan: New Styles and Old Patterns," *Ethnology* 41, no. 2 (Spring 2002): 119–134; and Murakami Motoko, "Eating Is a Solitary Pastime," *Japan Quarterly* 36 (April-June 1989): 207–210.

45. Alessandro Ferrara, ed., *Global Beef Trade* (New York: Nova Science, report sponsored by United States International Trade Commission, 2010), 48–60; Roxanne Clemens, "After the Ban: The Japanese Market for U.S. Beef," MATRIC Briefing Paper 07-MBP 12, Midwest Agribusiness Trade Research and Information Center, Iowa State University, Ames, Iowa, April 2007; "U.S. Opposes Japan's Request to Audit More U.S. Meatpacking Plants before Beef Trade Increases," *Food & Drink Weekly*, April 2, 2007, accessed January 15, 2009, www.thefreelibrary.com; "Innovative New Beef Cuts Boost U.S. Exports to Japan," *U.S. Federal News Service*, August 13, 2008; and Andy Sharp, "BSE and Foot-and-Mouth Disease Give Beef Producers Plenty to Chew on," *Japan Today*, September 2, 2010, accessed December 2, 2010, www.japantoday.com; Phyllis Entis, *Food Safety: Old Habits, New Perspectives* (Washington, DC: ASM Press, 2007), 202; and Matteo Ferrari, *Risk Perception, Culture, and Legal Change: A Comparative Study on Food Safety in the Wake of the Mad Cow Crisis* (Farnham, Surrey, UK: Ashgate, 2009), 70–75. For a comprehensive explanation of Japan's BSE safety mea-

sures, see Onodera Takashi and Chi-Kyeong Kim, "BSE Situation and Establishment of Food Safety Commission in Japan," *Journal of Veterinary Science* 7, no. 1 (March 2006): 1–11.

46. Clemens, "After the Ban"; Ken Y-N, "Most Japanese Choose McDonalds, Prefer MOS Burger," *What Japan Thinks*, October 3, 2007, whatjapanthinks.com; and "Real Value of Japan's One Trillion Yen BSE Tests," *The Cattle Site*, November 11, 2008, accessed January 10, 2009, www.thecattlesite.com.

47. "U.S. Beef Making Inroads in Japan," *The Cattle Site*, September 29, 2009, accessed December 4, 2009, www.thecattlesite.com; "Burger, Burger, Doughnut, Chicken," *What Japan Thinks*, October 22, 2009, accessed December 4, 2009, whatjapanthinks.com; Tabuchi Hiroko, "White House Presses Japan to Reopen Market to U.S. Beef," *New York Times*, April 7, 2010, accessed July 10, 2010, www.nytimes.com; "US-Japanese Beef Talks Continue," *The Beef Site*, April 8, 2010, accessed July 25, 2010, www.thebeefsite.com; Tabuchi Hiroko, "Disease Threatens Japan's Beef Trade," *New York Times*, June 11, 2010, accessed July 10, 2010, www.nytimes.com; "Japan Lifts FMD State of Emergency," *The Beef Site*, July 27, 2010, accessed September 15, 2011, www.thebeefsite.com; and "US Beef Imports Surge into Japan," *The Beef Site*, November 4, 2010, accessed December 2, 2010, www.thebeefsite.com.

48. Edward L. Rada, "Food Policy in China: Recent Efforts to Balance Supplies and Consumption Requirements," *Asian Survey* 23, no. 4 (1983): 520.

49. Ibid., 518–535; Buck, *Chinese Farm Economy*, 372; Buck, *Land Utilization in China*, 407; Kym Anderson and Rodney Tyers, "China's Economic Growth, Structural Transformation and Food Trade," *Australian Journal of Chinese Affairs* 14 (July 1985): 65–83.

50. Vaclav Smil, "Food Production and Quality of Diet in China," *Population and Development Review* 12, no. 1 (1986): 42; see also 25–45.

On the great famine of 1958 to 1962, see Frank Dikötter, *Mao's Great Famine: The History of China's Most Devastating Catastrophe, 1958–1962* (New York: Walker, 2012): in Hunan Province, the number of pigs dropped from 10.9 million in 1957 to 3.4 million in 1961 (141).

51. Ronald Coase and Ning Wang, *How China Became Capitalist* (New York: Palgrave Macmillan, 2012), 31, 46–48, 68; Peter J. Li, "Exponential Growth, Animal Welfare, Environmental and Food Safety Impact: The Case of China's Livestock Production," *Journal of Agricultural Environmental Ethics* 22 (2009): 230–231; Wei-Wei Zhang, *Transforming China: Economic Reform and Its Political Implications* (New York: St. Martin's, 2000), 9–10; Anderson and Tyers, "China's Economic Growth, Structural Transformation and Food Trade," 73–74; John W. Longworth, Colin G. Brown, and Scott A. Waldron, *Beef in China: Agribusiness Opportunities and Challenges* (St. Lucia, AU: University of Queensland Press, 2001), 240; MacLachlan, "Evolution of a Revolution," 34; Lester R. Brown, *Who Will Feed China? Wake-Up Call for a Small Planet* (New York: Norton, 1995), 46; and Barry Naughton, *The Chinese Economy: Transitions and Growth* (Cambridge: MIT Press, 2007), 242–266. Vaclav Smil estimated that China's average meat consumption went up by almost 300 percent between 1978 and 1994. See Smil, "Who Will Feed China?," *China Quarterly* 143 (September 1995): 809.

52. Robert F. Ash, "The Agricultural Sector in China: Performance and Policy Dilemmas during the 1990s," *China Quarterly* 131 (September 1992): 545–576; Smil, "Who Will Feed China?," 807; Barry M. Popkin, "Nutritional Patterns and Transitions," *Population and Development Review* 19, no. 1 (1993): 147–151; and Ross Garnaut and Guonan

Ma, "How Rich is China? Evidence from the Food Economy," *Australian Journal of Chinese Affairs* 30 (July 1993): 121–146.

53. Juday, "Intensification of Agriculture and Free Trade," 157–174; Peter J. Li, "Exponential Growth, Animal Welfare, Environmental and Food Safety Impact," 221–226, 234–235; MacLachlan, "Evolution of a Revolution," 34–35; Qiu Huajiao, Zhang Fusuo, Zhu Wanbin, Wang Haibin, and Cheng Xu, "Reorientation of China's Agriculture over the Next Two Decades," *Outlook on Agriculture* 37, no. 4 (2008): 250; and Brown, Longworth, and Waldron, *Regionalisation and Integration in China*, 8–13.

54. Scott A. Waldron, Colin Brown, John Longworth, and Zhang Cungen, *China's Livestock Revolution: Agribusiness and Policy Developments in the Sheep Meat Industry* (Cambridge, MA: CABI, 2007), 9, 55–68; Longworth, Brown, and Waldron, *Beef in China*, 30–32, 270; Brown, Longworth, and Waldron, *Regionalisation and Integration in China*, 16–29, 45–48, 53–66, 73; and Ferrara, *Global Beef Trade*, 86–87.

55. Waldron et al., *China's Livestock Revolution*, 56, 102–110; Longworth, Brown, and Waldron, *Beef in China*, 278–282; Brown, Longworth, and Waldron, *Regionalisation and Integration in China*, 49–52; Peter J. Li, "Exponential Growth, Animal Welfare, Environmental and Food Safety Impact," 235–237; MacLachlan, "Evolution of a Revolution," 35; and Huajiao et al., "Reorientation of China's Agriculture over the Next Two Decades," 251–253. On the issue of China's rapid dietary transition and health, see French and Crabbe, *Fat China*.

## Chapter 7

1. See, for instance, Yvonne Yen Liu, "Food Workers—Wages and Race," *Race, Poverty & the Environment* 18, no. 1 (2011): 10–12; and Alessandro Bonanno and Josefa Salete Barbosa Cavalcanti, eds., *Labor Relations in Globalized Food* (Bingley, UK: Emerald, 2014).

2. Sumii Sué, *The River with No Bridge*, trans. Susan Wilkinson (Tokyo: Tuttle, 1990; originally published in 1958), 348. On the issue of social status in the Tokugawa era in Japan, see David L. Howell, "Territoriality and Collective Identity in Tokugawa Japan," *Daedalus* 127, no. 3 (1998): 105–132. On the early history of the *Burakumin*'s establishment, see, for example, Suginohara, *Status Discrimination in Japan*; Neary, "*Burakumin* at the End of History," 269–294. Herbert Passin, "Untouchability in the Far East," *Monumenta Nipponica* 11, no. 3 (October 1955): 247–267; Kitaguchi, *An Introduction to the Buraku Issue*, 28. On the issue of the *eta* as a separate race, see Ninomiya Shigeaki, "An Inquiry Concerning the Origin, Development, and Present Situation of the *Eta* in Relation to the History of Social Classes in Japan," *Transactions of the Asiatic Society of Japan* 10 (December 1933): 49–153, esp. 54–59; and Smythe and Naitoh, "The Eta Caste in Japan," 19–20. The *paekchŏng*'s situation in Korea was quite similar to the *Buraku*. The *paekchŏng* existed as an outcaste group possibly as early as the tenth century but were certainly considered such by the thirteenth and fourteenth centuries. The ordinance of AD 968 had prohibited the slaughter of cattle but allowed the purchase and eating of it, though this was frowned upon. As Passin notes in his history of the *paekchŏng*, "somehow the feeling has always been strong that the slaughter and handling of meat is [*sic*] polluting and sinful, even if eating it is not." Passin, "The Paekchŏng of Korea: A Brief Social History," *Monumenta Nipponica* 12, no. 3/4 (1956–1957): 213. Also similar to the Japanese situation, the *paekchŏng* were divided historically into two groups: the *chaein*, which consisted primarily of actors, jugglers, acrobats, magicians, and "frivolous" el-

ements, and the *paekchŏng* proper, which consisted of butchers, leatherworkers, and basketmakers. In Korea's case, after the Mongol invasion during the 13th century, the *paekchŏng*'s status as meatworkers improved because of the Mongol's proclivities as meat eaters and their need for such labor.

3. John Locke, "Some Thoughts on Education," *The Works of John Locke*, 10th ed., vol. 6 (Darmstadt, DE: Scientia Verlag, 1963), 112–113.

4. David Perkins, *Romanticism and Animal Rights* (Cambridge: Cambridge University Press, 2003), 20–21; and Stephen F. Eisenman, *The Cry of Nature: Art and the Making of Animal Rights* (London: Reaktion Books, 2013), 15.

5. Kathryn Shevelow, *For the Love of Animals: The Rise of the Animal Protection Movement* (New York: Henry Holt, 2008), 5–13; Grier, *Pets in America*, 128–142; Vialles, *Animal to Edible*, 15–31.

6. James Plumptre, *The Experienced Butcher: Shewing the Respectability and Usefulness of His Calling, the Religious Considerations Arising from It, the Laws Relating to It and Various Profitable Suggestions for the Rightly Carrying It On: Designed Not Only for the Use of Butchers, but also for Families and Readers in General* (London: Darton, Harvey, and Darton; and Baldwin, Cradock, and Joy, 1816), 24–25.

7. Perkins, *Romanticism and Animal Rights*, 127–129; Charles Lamb, "Dissertation upon Roast Pig," in *The Complete Works and Letters of Charles Lamb* (New York: Modern Library, 1935), 108–113.

8. Gerald Carson, "The Great Meddler," *American Heritage* 19, no. 1 (December 1967): 28–33, 94–97; and Carson, *Men, Beasts, and Gods: A History of Cruelty and Kindness to Animals* (New York: Scribner, 1972), 129.

9. Ian MacLachlan, "Humanitarian Reform, Slaughter Technology, and Butcher Resistance in Nineteenth-Century Britain," in *Meat, Modernity, and the Rise of the Slaughterhouse*, ed. Paula Young Lee (Lebanon: University of New Hampshire Press, 2008), 112–117; and Roger Horowitz, *Kosher USA: How Coke Became Kosher and Other Tales of Modern Food* (New York: Columbia University Press, 2016), 167.

10. MacLachlan, "Humanitarian Reform, Slaughter Technology, and Butcher Resistance in Nineteenth-Century Britain," 115–117; Thomas Schlich, "The Word of God and the Word of Science: Nutrition Science and the Jewish Dietary Laws in Germany, 1820–1920," in *The Science and Culture of Nutrition, 1840–1940*, ed. Harmke Kamminga and Andrew Cunningham (Amsterdam: Rodopi B.V., 1995), 99–118; and Horowitz, *Kosher USA*, 212–215. In Germany, the debate over *shechita* started to gain intensity in the 1880s and 1890s. On the developments that led to Germany's as well as many other European countries' decisions to prohibit *shechita* between 1929 and 1940, see Brantz, "Stunning Bodies," 167–194.

11. Henry S. Salt, *Animals' Rights: Considered in Relation to Social Progress* (New York: Macmillan, 1894), 47–48; Peter Singer, *Animal Liberation*, 2nd ed. (New York: New York Review of Books, 1990); and Tom Regan, *The Case for Animal Rights* (Berkeley: University of California Press, 1983). In terms of contemporary positions on animal rights, perhaps the most advanced positions are arguments for recognition of legal rights for animals. See, for instance, Steve M. Wise, *Rattling the Cage: Toward Legal Rights for Animals* (Cambridge, MA: Perseus, 2000); and Wise, *Drawing the Line: Science and the Case for Animal Rights* (Cambridge, MA: Perseus Books, 2002).

12. Samuel Plimsoll, *Cattle Ships: The Fifth Chapter of Mr. Plimsoll's Second Appeal for our Seamen* (London: Kegan Paul, Trench, Trüber, 1890), 68, 74.

13. Salt, *Animals' Rights*, 48–49.

14. Ibid., 94–104.

15. H. F. Lester, *Behind the Scenes in Slaughter-Houses* (London: William Reeves, 1892), 5, 3–4, 4, 6, 6, 6, 5.

16. Smythe and Naitoh, "The Eta Caste in Japan," 23. Similarly, in Korea, official overturning of caste discrimination against the *paekchŏng* occurred in 1894, but bigotry continued. See William Shaw, "Between Class and Nation: The Equalization Society of the 1920s," in *Human Rights in Korea: Historical and Policy Perspectives*, ed. William Shaw (Cambridge, MA: East Asian Legal Studies Program of the Harvard Law School, 1991), 91–111.

17. Suginohara, *Status Discrimination in Japan*, chapters 4–5; and Neary, "*Burakumin* at the End of History," 270–271.

18. For a range of perspectives on the Japanese government's *Buraku* spending projects, see Nobuo Shimahara, "Toward the Equality of a Japanese Minority: The Case of *Burakumin*," *Comparative Education* 20, no. 3 (1984): 339–353; Kitaguchi, *An Introduction to the Buraku Issue*; and Neary, "*Burakumin* at the End of History." Neary estimates that $134 billion was spent on Dōwa projects; see 271.

19. Yagi, "Nihon no shokuniku," 14–17, 21, 28–29; and Kamata, *Documento Tojo*, 11–22, 101–105, 179–184.

20. Stephen H. Norwood, *Strikebreaking and Intimidation: Mercenaries and Masculinity in Twentieth-Century America* (Chapel Hill: University of North Carolina Press, 2002), 86–94.

21. Robert H. Zieger, *For Jobs and Freedom: Race and Labor in America since 1865* (Lexington: University Press of Kentucky, 2007), 58.

22. Ibid., 57–60. On the history of racial discrimination in Chicago's packinghouse industry before the 1930s, see James R. Barrett, *Work and Community in the Jungle: Chicago's Packinghouse Workers, 1894–1922* (Urbana: University of Illinois Press, 1987).

23. Norwood, *Strikebreaking and Intimidation*, 93. See also 86–94.

24. Rick Halpern, *Down on the Killing Floor: Black and White Workers in Chicago's Packinghouses, 1904–54* (Urbana: University of Illinois Press, 1997); Roger Horowitz, *"Negro and White, Unite and Fight!": A Social History of Industrial Unionism in Meatpacking, 1930–90* (Urbana: University of Illinois Press, 1997); Warren, *Tied to the Great Packing Machine*, 76–77; and Clayton Sinyai, *Schools of Democracy: A Political History of the American Labor Movement* (Ithaca, NY: ILR Press, 2006), 136–163.

25. Claflin, "La Villette," 34–39.

26. Oddy, *From Plain Fare to Fusion Food*, 16; and MacLachlan, "A Bloody Offal Nuisance," 249–254.

27. Roche, "International Food Regimes," 129–151; Fleming, "Agricultural Support Policies in a Small Open Economy," 334–354; Capie, "Australian and New Zealand Competition in the British Market, 1920–39," 46–63; Curtis and Reveley, "Producers, Processors and Unions," 135–158; Majorie A. Jerrard, "A Suprising Struggle? The AMIEU (Qld) and the Fight for Equal Wages in the Meat Processing and Export Industry in the 1950s and 1960s," *Labour History* 77 (1999): 140–159; Michael Barry, "A Bone of Contention: Managerial Initiative vs Employer Association Regulation of the New Zealand Meat Industry, 1960–75," *Labour History* 84 (2003): 69–88; and Cybèle Locke, *Workers in the Margins: Union Radicals in Post-War New Zealand* (Wellington, NZ: Bridgett Williams Books, 2012), 11, 18–19, 36–46, 64.

28. On TNCs' global labor sourcing strategies, see Philip McMichael, "The Power of Food," in *Globalization, Localization and Sustainable Livelihoods*, ed. Reidar Almås

and Geoffrey Lawrence (Aldershot, UK: Ashgate, 2003), 69–85; and McMichael, *Development and Social Change: A Global Perspective*, 3rd ed. (Thousand Oaks, CA: Pine Forge, 2004).

29. Warren, *Tied to the Great Packing Machine*, 23–26; J. Celeste Lay, *A Midwestern Mosaic: Immigration and Political Socialization in Rural America* (Philadelphia: Temple University Press, 2012), 13–15; William Kandel and Emilio A. Parrado, "Restructuring of the U.S. Meat Processing Industry and New Hispanic Migrant Destinations," *Population and Development Review* 31, no. 3 (2005): 447–471; LaGuana Gray, *We Just Keep Running the Line: Black Southern Women and the Poultry Processing Industry* (Baton Rouge: Louisiana State University Press, 2014), 127–130; Genoways, *The Chain*, 58; and Barry Estabrook, *Pig Tales: An Omnivore's Quest for Sustainable Meat* (New York: Norton, 2015), 214.

30. On ghettoization of modern meatpacking workers, see Michael Broadway and Donald Stull, "The Wages of Food Factories," *Food and Foodways* 18, no. 1/2 (2010): 48–49. Lay, in *A Midwestern Mosaic*, notes on page 2 that her study focuses on the process of how "native adolescents in ethnically diverse rural [meatpacking] communities begin to accept and welcome newcomers." For a detailed exploration of working and community interactions among African Americans and Latinos in the modern Southern rural poultry processing industry, see LaGuana Gray, *We Just Keep Running the Line*, especially 126–153. Other sources on the social isolation and discrimination against immigrants in the meat industry include Kathleen Stanley, "Industrial and Labor Market Transformation in the U.S. Meatpacking Industry," in *The Global Restructuring of Agro-Food Systems*, ed. Philip McMichael (Ithaca, NY: Cornell University Press, 1994), 129–144; Leon Fink, *The Maya of Morganton: Work and Community in the Nuevo New South* (Chapel Hill: University of North Carolina Press, 2003); Genoways, *The Chain*, 55–91, 162–172, 240–253; and Estabrook, *Pig Tales*, 211–213.

31. Locke, *Workers in the Margins*, 131, 156–164.

32. Gail A. Eisnitz, *Slaughterhouse: The Shocking Story of Greed, Neglect, and Inhumane Treatment inside the U.S. Meat Industry* (Amherst, NY: Prometheus, 1997), 188–189, 262–263.

33. Genoways, *The Chain*, 26–34, 255–259; Estabrook, *Pig Tales*, 191–199.

34. Pachirat, *Every Twelve Seconds*, 11.

35. "Migrant Seasonal Workers: The Impact on the Horticulture and Food Processing Sectors of Closing the Seasonal Agricultural Workers Scheme and the Sectors Based Scheme," Migration Advisory Committee, May 2013 (London: Migration Advisory Committee, 2013), 35, accessed July 17, 2014, www.ukba.homeoffice.gov.uk/mac.

36. Felicity Lawrence and Karen McVeigh, "'I'm Not a Slave, I Just Can't Speak English'—Life in the Meat Industry," *Guardian*, March 13, 2010, accessed July 17, 2014, www.theguardian.com/society/2010/mar/13/life-meat-industry-report-ehrc.

37. "Ireland: Immigrant Meat Workers," *Meat Trade News Daily*, April 16, 2011, accessed July 17, 2014, www.meattradenewsdaily.co.uk/news/290510/ireland_immigrants; Bettina Wagner and Anke Hassel, "Labor Migration and the German Meat Processing Industry: Fundamental Freedoms and the Influx of Cheap Labor," *South Atlantic Quarterly* 114, no. 1 (2015): 209–210; Deborah Gewertz and Frederick Errington, *Cheap Meat: Flap Food Nations in the Pacific Islands*, 40; and Keith Morgan, "Young Workers Super-Exploited in Australian Meat Industry," April 10, 2014, World Socialist Web Site, accessed July 17, 2014, www.wsws.org/en/2014/04/10/prim-a10.html; and Brown, Longworth, and Waldron, *Regionalisation and Integration in China*, 49.

38. Miura Koukichirou, "Tojyou wo Miru Me: Kouzouteki Sabetsu to Kankyou no Gensetsu no Aida" [Slaughterhouse in the eyes of the world: In between the structural discrimination and the discourse on environment], in *Sabetsu to Kankyo Mondai no Shakaigaku* [Sociology of discrimination and environmental problems], ed. Sakura Atsushi and Yoshii H. (Tokyo: Shinyosha, 2003).

39. Sakurai Atsushi, "Stories of the Slaughterhouse," *Jinbun Kenkyu* (*Journal of Humanities*, Faculty of Letters, Chiba University) 33 (2004): 71–77.

40. USDA, Economic Research Service, "Number of U.S. Farmers' Markets Continues to Rise," accessed June 19, 2016, www.ers.usda.gov/data-products/chart-gallery/detail.aspx?chartId=48561&ref=collection&embed=True; and Estabrook, *Pig Tales*, 235–277.

## Chapter 8

1. Although not directly related to the environmental problems stemming from increasing amounts of meat consumed around the globe, the proliferation of feral hogs, especially in the United States, Australia, and New Zealand, is another distinct environmental dilemma. Beginning in the 1980s, the number of wild hogs, which are part of the same species—*Sus scrofa*—as domesticated hogs, exploded in the United States and has reached at least six million today. In 2007, the estimated population of feral pigs in Australia was far higher—about 27 million, which made feral hogs more numerous than the country's population of 21 million. Although they will attack lambs and calves and cause damage by trampling and feeding on crops and wrecking fences and waterholes, their greatest threat is as potential carriers of several epizootic diseases that could potentially affect commercial swine, including pseudorabies, brucellosis, classical swine fever (hog cholera), and foot-and-mouth disease. Foot-and-mouth disease, in particular, can also be spread to cattle and sheep, and most experts consider it to be the most significant threat to international livestock and meat trade in the world today. Estimates of the potential lost revenue due to a disease outbreak among domestic hogs are about $4 billion in the United States and $3 to $5 billion in Australia. On the history and current problems of feral pigs in the United States, see, for example, John J. Mayer and I. Lehr Brisbin, Jr., *Wild Pigs in the United States: Their History, Comparative Morphology, and Current Status* (Athens: University of Georgia Press, 1991), 2–3, 24, 34, 40, 43–44, 46–47, 226; Michael D. Thompson, "This Little Piggy Went to Market: The Commercialization of Hog Production in Eastern North Carolina from William Shy to Wendell Murphy," *Agricultural History* 74, no. 2 (2000): 580–581; Estabrook, *Pig Tales*, 48–60; Matthew Sculley, *Dominion: The Power of Man, the Suffering of Animals, and the Call to Mercy* (New York: St. Martin's, 2002), 236–238, 263–277; and John Morthland, "A Plague of Pigs in Texas," *Smithsonian* 41, no. 9 (2011): 52–61. On Australia's feral pig issues, see, for instance, George Wilson, Nick Dexter, Peter O'Brien, and Mary Bomford, *Pest Animals in Australia: A Survey of Introduced Wild Animals* (Kenthurst, AU: Bureau of Rural Resources and Kangaroo Press, 1992), 5, 46; Penny Olsen, *Australia's Pest Animals: New Solutions to Old Problems* (Canberra: Bureau of Resource Sciences, Commonwealth of Australia, 1998), 13, 16, 24, 36, 38, 61, 81, 121; and Tim Low, *Feral Future: The Untold Story of Australia's Exotic Invaders* (Chicago: University of Chicago Press, 2002; originally published in 1999), 230–231, 247.

2. Richard Jones, "Why Manure Matters," in *Manure Matters: Historical, Archaeological and Ethnographic Perspectives*, ed. Richard Jones (Farnham, Surrey, UK: Ashgate, 2012), 1–11.

3. King, *Farmers of Forty Centuries*, 193; see also 200.

4. Albert Howard and Yeshwant D. Wad, *The Waste Products of Agriculture: Their Utilization as Humus* (London: Humphrey Milford and Oxford University Press, 1931), 41, 60; J. Ronald Miner, Frank J. Humenik, and Michael R. Overcash, *Managing Livestock Wastes to Preserve Environmental Quality* (Ames: Iowa State University Press, 2000), 203; J. Thomas Sims, "Characteristics of Animal Wastes and Waste-Amended Soils: An Overview of the Agricultural and Environmental Issues," in *Animal Waste and the Land-Water Interface*, ed. Kenneth Steele (Boca Raton, FL: CRC, 1995), 1–2; MacLachlan, "Evolution of a Revolution," 32. On the larger implications of the transition to CAFO-based agriculture in the United States, see Terence J. Centner, *Empty Pastures: Confined Animals and the Transformation of the Rural Landscape* (Urbana: University of Illinois Press, 2004).

5. Day, "Butchers, Tanners, and Tallow Chandlers," 81–102; MacLachlan, "A Bloody Offal Nuisance," 233–235, 241–242; Robyn S. Metcalfe, *Meat, Commerce and the City: The London Food Market, 1800–1855* (London: Pickering and Chatto, 2012), 49–108; Christine Meisner Rosen, "The Role of Pollution Regulation and Litigation in the Development of the U.S. Meatpacking Industry, 1865–1880," *Enterprise & Society* 8, no. 2 (2007): 297–347; Andrew Hurley, "Busby's Stink Boat and the Regulation of Nuisance Trades, 1865–1918," 145–162; Shirley Fisher, "The Pastoral Interest and Sydney's Public Health," 84–85; and Sakurai, "Stories of the Slaughterhouse," 81.

6. Warren, *Tied to the Great Packing Machine*, 172–176.

7. Miner, Humenik, and Overcash, *Managing Livestock Wastes to Preserve Environmental Quality*, 3, 27–29; Joseph P. Zublena, "Characteristics of Animal Wastes and Waste-Amended Soils: Nutrient Management Planning beyond the Farm Boundary," in *Animal Waste and the Land-Water Interface*, ed. Kenneth Steele (Boca Raton, FL: CRC, 1995), 49–50; and Philip W. Gassman and Aziz Bouzaher, "Livestock Pollution: Lessons from the European Union," in *Animal Waste and the Land-Water Interface*, ed. Kenneth Steele (Boca Raton, FL: CRC, 1995), 520.

8. Miner, Humenik, and Overcash, *Managing Livestock Wastes to Preserve Environmental Quality*, 108–126; Robert L. Mikkelsen and J. Wendell Gilliam, "Animal Waste Management and Edge of Field Losses," in *Animal Waste and the Land-Water Interface*, ed. Kenneth Steele (Boca Raton, FL: CRC, 1995), 58–61; L. S. Shore, D. L. Correll, and P. K. Chakraborty, "Relationship of Fertilization with Chicken Manure and Concentrations of Estrogens in Small Streams," in *Animal Waste and the Land-Water Interface*, ed. Kenneth Steele (Boca Raton, FL: CRC, 1995), 157; and Vaclav Smil, *Enriching the Earth: Fritz Haber, Carl Bosch, and the Transformation of World Food Production* (Cambridge: MIT Press, 2001), 188–192.

9. Miner, Humenik, and Overcash, *Managing Livestock Wastes to Preserve Environmental Quality*, 150–164, 170–180, 241; A. Breeuwsma, J. G. A. Reijerink, and O. F. Schoumans, "Impact of Manure on Accumulation and Leaching of Phosphate in Areas of Intensive Livestock Farming," in *Animal Waste and the Land-Water Interface*, ed. Kenneth Steele (Boca Raton, FL: CRC, 1995), 239; Gassman and Bouzaher, "Livestock Pollution," 515; Nigel Key, William D. McBride, and Marc Ribaudo, *Changes in Manure Management in the Hog Sector: 1998–2004* (USDA, Economic Research Service, Economic Information Bulletin Report No. 50, March 2009), 5–6, 9; Nigel Key, William D. McBride, Marc Ribaudo, and Stacy Sweeringer, *Trends and Developments in Hog Manure Management: 1998–2009* (USDA, Economic Research Service, Economic Information Bulletin Report No. 81, September 2011), 3–6, 23–24; and Estabrook, *Pig Tales*, 118–122.

10. Elisabeth A. Stoddard, "Neoliberal Governance and Environmental Risk: 'Normal Accidents' in North Carolina's Hog Industry," in *Political Ecologies of Meat*, ed. Jody Emel and Harvey Neo (London: Routledge, 2015), 137–158; Nicolette Hahn Niman, *Righteous Pork Chop: Finding a Life and Good Food beyond Factory Farms* (New York: Collins Living, 2009), 11–24; Thomas J. Hoban and William B. Clifford, "Livestock and Poultry Producers' Waste Management Practices and Attitudes in North Carolina," in *Animal Waste and the Land-Water Interface*, ed. Kenneth Steele (Boca Raton, FL: CRC, 1995), 446.

11. David Saurí and Hug March, "Can't Go to the Fountain No More: Pigs, Nitrates and Spring Water Pollution in Catalonia," in *Political Ecologies of Meat*, ed. Jody Emel and Harvey Neo (London: Routledge, 2015), 109–126; Gassman and Bouzaher, "Livestock Pollution," 515–520; Tiffany T. Y. Guan and Richard A. Holley, *Hog Manure Management, the Environment and Human Health* (New York: Kluwer Academic/ Plenum, 2003), 123–125; Miner, Humenik, and Overcash, *Managing Livestock Wastes to Preserve Environmental Quality*, 241; Clare Narrod and Carl Pray, "Technology Transfer in the Poultry Industry: Factors Associated with Increased Production," in *Animal Waste and the Land-Water Interface*, ed. Kenneth Steele (Boca Raton, FL: CRC, 1995), 529–530; MacLachlan, "Evolution of a Revolution," 34–36; and Smil, *Should We Eat Meat?*, 121.

12. Dennis F. Whigman, "The Role of Wetlands, Ponds, and Shallow Lakes in Improving Water Quality," in *Animal Wastes and the Land-Water Interface*, ed. Jody Emel and Harvey Neo (London: Routledge, 2015), 169; Robert H. Kadlec, "Design Models for Nutrient Removal in Constructed Wetlands," in *Animal Wastes and the Land-Water Interface*, ed. Jody Emel and Harvey Neo (London: Routledge, 2015), 173, 182–183; Guan and Holley, *Hog Manure Management*, 147–148, 151–152; Key, McBride, and Ribaudo, *Changes in Manure Management in the Hog Sector, 1998–2004*, 7–8, 16–19; Key, McBride, Ribaudo, and Sweeringer, *Trends and Developments in Hog Manure Management: 1998–2009*, 8–9, 24–25; Miner, Humenik, and Overcash, *Managing Livestock Wastes to Preserve Environmental Quality*, 187, 251; Martha Noble, "Paying the Polluters: Animal Factories Feast on Taxpayer Subsidies," in *The CAFO Reader: The Tragedy of Industrial Animal Factories*, ed. Daniel Imhoff (Berkeley, CA: Watershed Media, 2010), 221–226; and Paige Tomaselli and Meredith Niles, "Changing the Law: The Road to Reform," in *The CAFO Reader: The Tragedy of Industrial Animal Factories*, ed. Daniel Imhoff (Berkeley, CA: Watershed Media, 2010), 316–321, 327.

13. Niman, *Righteous Pork Chop*, 155–156; Miner, Humenik, and Overcash, *Managing Livestock Wastes to Preserve Environmental Quality*, 242; Gene Logsdon, *Holy Shit: Managing Manure to Save Mankind* (White River Junction, VT: Chelsea Green, 2010), 35; T. A. McCaskey, "Feeding Broiler Poultry Litter as an Alternative Waste Management Strategy," in *Animal Waste and the Land-Water Interface*, ed. Kenneth Steele (Boca Raton, FL: CRC, 1995), 493–502; and Phillip C. Badger, Janice K. Lindsey, and John D. Veitch, "Energy Production from Animal Wastes," in *Animal Waste and the Land-Water Interface*, ed. Kenneth Steele (Boca Raton, FL: CRC, 1995), 480.

14. Miner, Humenik, and Overcash, *Managing Livestock Wastes to Preserve Environmental Quality*, 23–24, 38, 257–258, 265, 273; Niman, *Righteous Pork Chop*, 16; Warren, *Tied to the Great Packing Machine*, 166–172; Genoways, *The Chain*, 205; and Estabrook, *Pig Tales*, 115.

15. Henning Steinfield, Pierre Gerber, Tom Wassenaar, Vincent Castel, Mauricio Rosales, and Cees de Haan, *Livestock's Long Shadow: Environmental Issues and Options* (Rome: Food and Agriculture Organization of the United Nations, 2006).

16. Simon Fairlie, *Meat: A Benign Extravagance* (White River Junction, VT: Chelsea Green, 2010), 170; see also 157–183.

17. Francis M. Kelliher and Harry Clark, "Ruminants," in *Methane and Climate Change*, ed. Dave Reay, Pete Smith, and André van Amstel (London: Earthscan, 2010), 136; Smil, *Enriching the Earth*, 182; Philip K. Thornton and Pierre J. Gerber, "Climate Change and the Growth of the Livestock Sector in Developing Countries," *Mitigation and Adaptation Strategies for Global Change* 15 (2010): 175; Tasneem Abbasi, S. M. Tauseef, S. A. Abbasi, *Biogas Energy* (New York: Springer, 2012), 25–28; Weis, *The Ecological Hoofprint*, 134; Dave Reay, Pete Smith, and André van Amstel, "Methane Sources and the Global Methane Budget," in *Methane and Climate Change*, ed. Dave Reay, Pete Smith, and André van Amstel (London: Earthscan, 2010), 5; and Smil, *Should We Eat Meat?*, 169–170.

18. Keith Smith, Introduction, in *Nitrous Oxide and Climate Change*, ed. Keith Smith (London: Earthscan, 2010), 1–2; Cecile de Klein, Richard Eckard, and Tony van der Weerden, "Nitrous Oxide Emissions from the Nitrogen Cycle in Livestock Agriculture: Estimation and Mitigation," in *Nitrous Oxide and Climate Change*, ed. Keith Smith (London: Earthscan, 2010), 107–142; Keith C. L. Lee, Joshua P. Newell, Jennifer R. Wolch, and Pascale Joassart-Marcelli, "Cows, Climate and the Media," in *Political Ecologies of Meat*, ed. Jody Emel and Harvey Neo (London: Routledge, 2015), 195–215; and Smil, *Should We Eat Meat?*, 171.

19. Pacheco and Poccard-Chapuis, "Cattle Ranching Development in the Brazilian Amazon," 42–66; and Fairlie, *Meat*, 163–165.

20. *Greenhouse Gas Emissions from the Australian Beef and Sheep Meat Industries*, M245, Bureau of Resource Science, Meat & Livestock, Australia, Ltd., July 1994, 1–5, 11–12, 18–29, 37, 62–65; *Environmental Factors Affecting Australia's Livestock Industries*, Bureau of Rural Sciences, Commonwealth Department of Agriculture, Fisheries and Forestry of Australia, Kingston, Australia, June 2001, 27–30; Thornton and Gerber, "Climate Change and the Growth of the Livestock Sector in Developing Countries," 175–176; and Christopher Rosin and Mark H. Cooper, "Mitigating Greenhouse Gas Emissions from Livestock: Complications, Implication and New Political Ecologies," in *Political Ecologies of Meat*, ed. Jody Emel and Harvey Neo (London: Routledge, 2015), 315–328.

21. Tomaselli and Niles, "Changing the Law: The Road to Reform," 314–329; Guan and Holley, *Hog Manure Management*, 151; Key, McBride, and Ribaudo, *Changes in Manure Management in the Hog Sector, 1998–2004*, 7; Key, McBride, Ribaudo, and Sweeringer, *Trends and Developments in Hog Manure Management: 1998–2009*, 9; and Claudia Copeland, "Air Quality Issues and Animal Agriculture: A Primer," *Congressional Research Service*, 7-5700, RL32948, December 22, 2014, 21–26.

22. Smil, *Enriching the Earth*, 205; and Thornton and Gerber, "Climate Change and the Growth of the Livestock Sector in Developing Countries," 177–180.

23. Abbasi, Tauseef, and Abbasi, *Biogas Energy*, 12–14, 18, 50–54, 118; and Miriam H. A. van Eekert, Hendrik Jan van Dooren, Marjo Lexmond, and Grietje Zeeman, "Wastewater and Manure," in *Methane and Climate Change*, ed. Dave Reay, Pete Smith, and André van Amstel (London: Earthscan, 2010), 151–174.

## Chapter 9

1. The quote is originally from 1868 but is cited here from Ellen Gould Harmon White, *Counsels on Health and Instruction to Medical Missionary Workers* (Mountain View, CA: Pacific, 1923), 576.

2. James C. Riley, "Height, Nutrition, and Mortality Risk Reconsidered," *Journal of Interdisciplinary History* 24, no. 3 (1994): 470; Ina Zweiniger-Bargielowska, "*The Culture of the Abdomen*: Obesity and Reducing in Britain, circa 1900–1939," *Journal of British Studies* 44, no. 2 (2005): 252, 268; Patricia Hausman, *Jack Sprat's Legacy: The Science and Politics of Fat and Cholesterol* (New York: Richard Marek, 1981), 55; Laura S. Sims, *The Politics of Fat: Food and Nutrition Policy in America* (Armonk, NY: M. E. Sharpe, 1998), 50–51; United States Senate, *Hearings before the Select Committee on Nutrition and Human Needs*, 95th Congress, 1st Session, March 24, 1977 (Washington, DC: GPO, 1977), 10; Marion Nestle, *Food Politics: How the Food Industry Influences Nutrition and Health* (Berkeley: University of California Press, 2002), 39.

3. Keys and Keys, *Eat Well and Stay Well*, 42. On the significance of the Keyses' work, also see Smil, *Feeding the World*, 268.

4. Hausman, *Jack Sprat's Legacy*, 28–33, 96–100, 148–150; Sims, *The Politics of Fat*, 266; Jessica J. Mudry, *Measured Meals: Nutrition in America* (Albany: State University of New York Press, 2009), 25–43, 47–68; and Smil, *Feeding the World*, 268–272.

5. George McGovern, *The Third Freedom: Ending Hunger in Our Time* (New York: Simon and Schuster, 2001), 76; see also 70–76.

6. *Dietary Goals for the United States*, 2nd ed., xxi.

7. Nestle, *Food Politics*, 38–40.

8. United States Senate, *Hearings before the Select Committee on Nutrition and Human Needs*, March 24, 1977, 4, 26.

9. Hausman, *Jack Sprat's Legacy*, 31.

10. *Dietary Goals for the United States*, 4, 43, 54. See also Mudry, *Measured Meals*, 40, 82.

11. Hausman, *Jack Sprat's Legacy*, 38, 178–184.

12. Sims, *The Politics of Fat*, 233; see also 173, 233–234.

13. Mudry, *Measured Meals*, 77, 83–84, 88–95, 105–109; Nestle, *Food Politics*, 40–50.

14. "IARC Monographs Evaluate Consumption of Red Meat and Processed Meat," October 26, 2015, 1, accessed December 15, 2015, www.iarc.fr/en/media-centre/pr/2015/pdfs/pr240_E.pdf.

15. Sims, *The Politics of Fat*, 15; Charlotte Biltekoff, *Eating Right in America* (Durham, NC: Duke University Press, 2013), 117–118; "Is Total Fat Consumption Really Decreasing?," *USDA Center for Nutrition Policy and Promotion*, April 1998, accessed November 18, 2015, www.cnpp.usda.gov/sites/default/files/nutrition_insights_uploads/insight5.pdf; and Vaclav Smil, "Coronary Heart Disease, Diet, and Western Mortality," *Population and Development Review* 15, no. 3 (September 1989): 402, 412. On the worldwide downturn in beef consumption and the rise of poultry due partly to health concerns during the 1970s, see Perren, *Taste, Trade and Technology*, 195–208.

16. Smil, *Should We Eat Meat?*, 24–25. On the 1980s as a turning point in global consumers' concerns about meat safety, see Perren, *Taste, Trade and Technology*, 228.

17. Ian C. Shaw, *Food Safety: The Science of Keeping Food Safe* (West Sussex, UK: Wiley-Blackwell, 2013), 141; T. Hugh Pennington, *When Food Kills: BSE, E. Coli, and Disaster Science* (Oxford: Oxford University Press, 2003), 124–126; Entis, *Food Safety*, 179–181; and Claude Fischler, "The 'Mad Cow' Crisis: A Global Perspective," in *Food in Global History*, ed. Raymond Grew (Boulder, CO: Westview, 1999), 207–213.

18. Christopher D. Cook, *Diet for a Dead Planet: How the Food Industry Is Killing Us* (New York: New Press, 2004), 28–29; Entis, *Food Safety*, 181–183, 185–189, 198; Pen-

nington, *When Food Kills*, 147–164; Perren, *Taste, Trade and Technology*, 229–230; and Marion Nestle, *Safe Food: Bacteria, Biotechnology, and Bioterrorism* (Berkeley: University of California Press, 2003), 251–252.

19. Entis, *Food Safety*, 183, 197–199; Maxime Schwartz, *How the Cows Turned Mad*, trans. Edward Schneider (Berkeley: University of California Press, 2003), 72, 119, 142–153; Perren, *Taste, Trade and Technology*, 230–232; Nestle, *Safe Food*, 252–253; and Ferrari, *Risk Perception*, 48.

20. Ferrari, *Risk Perception*, 17, 36, 49, 59–63, 75–80; Karolina Zurek, *European Food Regulation after Enlargement: Facing the Challenges of Diversity* (Leiden, NL: Martinus Nijhoff, 2012), 111–148; Entis, *Food Safety*, 113; Ian C. Shaw, *Food Safety*, 147–151; and Cook, *Diet for a Dead Planet*, 29. Also see Codex Alimentarius International Food Standards website: www.fao.org/fao-who-codexalimentarius/en/.

21. Cook, *Diet for a Dead Planet*, 33–35; Entis, *Food Safety*, 62, 113, 202–207; Ian C. Shaw, *Food Safety*, 154; Perren, *Taste, Trade and Technology*, 234–235; Nestle, *Safe Food*, 70–72, 251–255; Ferrari, *Risk Perception*, 67–69; and Genoways, *The Chain*, 26–34.

22. Pennington, *When Food Kills*, 97–104; Entis, *Food Safety*, 72–73, 89–90, 97, 113, 119, 125–126; Ian C. Shaw, *Food Safety*, 78–79; Nestle, *Safe Food*, 86–112; and Smil, *Should We Eat Meat?*, 25.

23. Nestle, *Safe Food*, 36; Ian C. Shaw, *Food Safety*, 66–100, 109, 130–138; Anastacia Marx de Salcedo, *Combat-Ready Kitchen: How the U.S. Military Shapes the Way You Eat* (New York: Current, 2015), 62–64; Mark Bittman, *A Bone to Pick: The Good and Bad News about Food, with Wisdom and Advice on Diets, Food Safety, GMOs, Farming, and More* (New York: Pam Krauss, 2015), 57; Charles M. Duncan, *Eat, Drink, and Be Wary: How Unsafe Is Our Food?* (Lanham, MD: Rowman and Littlefield, 2015), 6; Smil, *Should We Eat Meat?*, 26; and Estabrook, *Pig Tales*, 164–165.

24. On the numbers of food-related illness in the United States, see Jeffrey Burkhardt, "The Ethics of Food Safety in the Twenty-first Century: Who Keeps the Public Good," in *The Philosophy of Food*, ed. David M. Kaplan (Berkeley: University of California Press, 2012), 141. On the use of the term *flexitarian*, see Rod Preece, *Sins of the Flesh: A History of Vegetarian Thought* (Vancouver: UBC, 2008), 15.

25. For estimates of the numbers of vegetarians in North America and Europe during the first decade of the twenty-first century, see Preece, *Sins of the Flesh*, 2; Donna Maurer, *Vegetarianism: Movement or Moment?* (Philadelphia: Temple University Press, 2002), 15–17, 144; Peter Singer and Jim Mason, *The Way We Eat: Why Our Food Choices Matter* (Emmaus, PA: Rodale, 2006), 187; Zaraska, *Meathooked*, 136–146; and Laura K. Hahn, "I'm Too Sexy for Your Movement: An Analysis of the Failure of the Animal Rights Movement to Promote Vegetarianism," in *Arguments about Animal Ethics*, ed. Greg Goodale and Jason Edward Black (Lanham, MD: Lexington, 2010), 81–82. On comparisons of the size of the nineteenth-century Western vegetarian movement to the present, see Perren, *Taste, Trade and Technology*, 190–193. Ian Watson, one of the best-known British science fiction writers, wrote the novel *Meat* in 1988 as the BSE scare and resulting meat avoidance movements gained momentum.

26. On the religious roots of nineteenth-century British and American vegetarianism, see James Gregory, *Of Victorians and Vegetarians*, 100–108; Preece, *Sins of the Flesh*, 270; and Colin Spencer, *The Heretic's Feast: A History of Vegetarianism* (Hanover, NH: University Press of New England, 1995). On vegetarianism in South and East Asia, see Ludwig Alsdorf, *The History of Vegetarianism and Cow-Veneration in India*, ed. Willem Bollée, trans. Bal Patil (London: Routledge, 2010; originally published in 1962), 1–67;

James Stewart, *Vegetarianism and Animal Ethics in Contemporary Buddhism* (London: Routledge, 2016), 2–3, 14–15, 27–28, 48–49, 78, 123–126, 176, 185; Colleen Taylor Sen, "Jainism: The World's Most Ethical Religion," in *Food and Morality*, ed. Susan R. Friedland (Devon, UK: Prospect, 2008), 230–240; Preece, *Sins of the Flesh*, 3; and Zaraska, *Meathooked*, 168–182.

27. Preece, *Sins of the Flesh*, 18, 304–307; Maurer, *Vegetarianism*, 3–20, 30–45; Singer and Mason, *The Way We Eat*, 224–228; and Gary E. Fraser, *Diet, Life Expectancy, and Chronic Disease: Studies of Seventh-Day Adventists and Other Vegetarians* (Oxford: Oxford University Press, 2003), 251.

28. Preece, *Sins of the Flesh*, 16, 323; Maurer, *Vegetarianism*, 5, 50, 78; Singer and Mason, *The Way We Eat*, 187; Gary L. Francione, "Animal Welfare, Happy Meat, and Veganism as the Moral Baseline," in *Philosophy of Food*, ed. David M. Kaplan (Berkeley: University of California Press, 2012), 169–189; and Steven Best, *The Politics of Total Liberation: Revolution for the 21st Century* (New York: Palgrave Macmillan, 2014). One example of a particularly ardent vegan attack on traditional food culture and assertions about the strength-building notions associated with meat eating is John Joseph's *Meat Is for Pussies: A How-To Guide for Dudes Who Want to Get Fit, Kick Ass, and Take Names* (New York: HarperWave, 2014). For a critical assessment of veganism as an aspiration rather than a realistic attainment, see Lori Gruen and Robert C. Jones, "Veganism as an Aspiration," in *The Moral Complexities of Eating Meat*, ed. Ben Bramble and Bob Fischer (Oxford: Oxford University Press, 2016), 153–171.

29. On animal by-product use in the contemporary United States, see Herbert W. Ockerman and C. L. Hansen, *Animal By-Product Processing and Utilization* (Lancaster, PA: Technomic, 2000).

30. Francione, "Animal Welfare, Happy Meat, and Veganism as the Moral Baseline," 169–189; Hahn, "I'm Too Sexy for Your Movement," 79–95; and James McWilliams, *The Modern Savage*.

31. On the benefits of diets based on perennial crop polycultures, see Michael Pollan, *The Omnivore's Dilemma: A Natural History of Four Meals* (New York: Penguin, 2006); and Lierre Keith, *The Vegetarian Myth: Food, Justice, and Sustainability* (Crescent City, CA: Flashpoint, 2009). On the benefits of a Paleolithic diet, see Staffan Lindeberg, *Food and Western Disease: Health and Nutrition from an Evolutionary Perspective* (Chichester, West Sussex, UK: Wiley-Blackwell, 2010); and Le, *100 Million Years of Food*, 55–58. For a critique of the current US food guidelines, see Gary Taubes, "Good Calories, Bad Calories: What Really Makes Us Fat," *Mother Earth News* 230 (October/November 2008): 61–64.

32. Perren, *Taste, Trade and Technology*, 205–208.

33. Keys and Keys, *Eat Well and Stay Well*, 154.

34. *Dietary Goals for the United States*, 2nd ed., 4.

35. Dixon, *Changing Chicken*, 71–79; Striffler, *Chicken*, 15, 27–30; and Schatzker, *The Dorito Effect*, 21–22, 26–29, 33–40.

36. "What Is 'Shokuiku (Food Education)'?," accessed December 12, 2015, www.maff.go.jp/e/pdf/shokuiku.pdf; and Abdul Jalil Rohana, Naomi Aiba, Nobuo Yoshiike, and Miki Miyoshi, "Japan for Sustainability in Health through a New Movement of Food and Nutrition Education 'Shokuiku,'" *International Medical Journal* 18, no. 1 (2011): 21–28. Also see Justin McCurry, "Japan Battles with Obesity," *Lancet* 369: 9560 (2007): 451–453.

37. Smil and Kobayashi, *Japan's Dietary Transition and Its Impacts*, 95; see also 13, 88.

38. On the importance of animal protein in increasing height in various national populations in the twentieth century, see Joerg Baten and Matthias Blum, "Why Are You Tall While Others Are Short? Agricultural Production and Other Proximate Determinants of Global Heights," *European Review of Economic History* 18 (2014): 144–165. In *100 Million Years of Food*, Stephen Le emphasizes that "[e]ating more animal foods will probably make you taller, stronger, more fertile, and feel better, but these foods will also likely shorten your life span" (5).

39. "What Is 'Shokuiku (Food Education)'?," 3.

40. Ibid., 8.

41. Ibid., 2.

42. T. Colin Campbell and Thomas M. Campbell, *The China Study: The Most Comprehensive Study of Nutrition Ever Conducted and the Startling Implications for Diet, Weight Loss and Long-Term Health* (Dallas: Benbella, 2006), 7.

43. Ibid., 3–7, 243.

44. French and Crabbe, *Fat China*; and "Chronic Disease on Rise in China: Health Survey," June 30, 2015, accessed December 14, 2015, news.xinhuanet.com/English/2015-06/30/c_134369371.htm.

## Chapter 10

1. Smil, *Should We Eat Meat?*, 205.

2. Ibid., 178–215. Also see Robert Paarlberg, *Food Politics: What Everyone Needs to Know* (Oxford: Oxford University Press, 2010).

3. Nikos Alexandratos and Jelle Bruinsma, "World Agriculture towards 2030/2050: The 2012 Revision," ESA Working Paper No. 12–03, June 2012, Food and Agriculture Organization of the United Nations, accessed December 28, 2015, www.fao.org/docrep/016/ap106e/ap106e.pdf; Colin Tudge, "How to Raise Livestock—And How Not To," in *The Meat Crisis: Developing More Sustainable Production and Consumption*, ed. Joyce D'Silva and John Webster (London: Earthscan, 2010), 15–16.

4. Smil, *Should We Eat Meat?*, 215. Also see Smil, *Feeding the World*, 141–180.

5. Nicholas Kristof, "The (Fake) Meat Revolution," *New York Times*, September 15, 2015, accessed July 11, 2016, www.nytimes.com/2015/09/20/opinion/sunday/nicholas-kristof-the-fake-meat-revolution.html. On entomophagy, see Zaraska, *Meathooked*, 185–190; Arnold van Huis, Joost Van Itterbeeck, Harmke Klunder, Esther Mertens, Afton Halloran, Giulia Muir, and Paul Vantomme, *Edible Insects: Future Prospects for Food and Feed Security* (Rome: Food and Agriculture Organization of the United Nations, 2013), 1, 15–18, 35–41, 59, 67, 74–75, 89–90, 107, 108–111.

6. Daniella Martin, *Edible: An Adventure into the World of Eating Insects and the Last Great Hope to Save the Planet* (Boston: New Harvest, 2014), 113.

7. Van Huis et al., *Edible Insects*, 154–156.

8. Niina Heikkinen, "A Carbon Tax on Meat?," *Scientific American*, ClimateWire, February 4, 2016, accessed July 11, 2016, www.scientificamerican.com/article/a-carbon-tax-on-meat/; Adam Withnall, "Denmark Ethics Council Calls for Tax on Red Meat to Fight 'Ethical Problem' of Climate Change," *Independent*, April 27, 2016, accessed July 11, 2016, www.independent.co.uk/news/world/europe/denmark-ethics-council-calls-for-tax-on-red-meat-to-fight-ethical-problem-of-climate-change-a7003061.html; and Zaraska, *Meathooked*, 192.

9. The Brundtland Commission report (accessed December 28, 2015) is located at www.un-documents.net/our-common-future.pdf.

10. Joan Dye Gussow, "Ecology and Vegetarian Considerations: Does Environmental Responsibility Demand the Elimination of Livestock?," *American Journal of Clinical Nutrition* 59, no. 5 (1994): 1115S; see also 1110S–1116S.

11. On the importance of a balanced nutritional view of meat's role in human lives, see Smil, *Should We Eat Meat?*, 178–215; and Johanna T. Dwyer, "Vegetarian Eating Patterns: Science, Values, and Food Choices—Where Do We Go from Here?," *American Journal of Clinical Nutrition* 59, no. 5 (1994): 1255S–1262S.

12. Plumptre, *The Experienced Butcher*, 20, 8.

13. Perkins, *Romanticism and Animal Rights*, 31.

14. Ruth Harrison, *Animal Machines: An Exposé of "Factory Farming" and Its Dangers to the Public*, 2nd ed. (New York: Ballantine, 1966), 17.

15. Kate Rawles, "Developing Ethical, Sustainable and Compassionate Food Policies," in *The Meat Crisis: Developing More Sustainable Production and Consumption*, ed. Joyce D'Silva and John Webster (London: Earthscan, 2010), 215, 222. On the Farm Animal Welfare Committee, see http://www.gov.uk/government/groups/farm-animal-welfare-committee-fawc, accessed December 28, 2015. The animal liberation movement, however, quickly divided over the issue of speciesism, a term that Peter Singer used to explain the prejudice toward sentient creatures that allows humans to hurt, kill, and consume nonhuman animals. Singer strongly insisted in *Animal Liberation* that animals have the right to not be treated in a way that causes them pain. He also urged avoidance of meat eating, yet found either vegetarianism or veganism acceptable. Tom Regan built on Singer's work but developed a moral philosophy about animal rights that rests on animals' consciousness or self-awareness, a field that has grown considerably since Regan's *The Case for Animal Rights* was published in 1983. In Regan's assessment, animals—at least those with consciousness or self-awareness—deserve equal rights with humans. Regarding agricultural practices, whereas Singer urged the end of industrial, intensive livestock practices, Regan goes even farther. People "should not be satisfied with anything less than the total dissolution of commercial animal agriculture as we know it, whether factory farms or otherwise"; *The Case for Animal Rights*, 351. R. G. Frey and Alan White, for instance, disagree with Regan, noting that animals do not have beliefs and do not have the ability to exercise, own, or enjoy rights. Animal liberationists also differ over the use of violence to end meat consumption. Bob Torres says violence would be counterproductive while Stephen Best says animal liberation demands direct action and radical transformations. See Singer, *Animal Liberation*, esp. 8–9, 21, 69, 83, 159–161, 177, 226; Regan, *Animal Rights, Animal Wrongs: An Introduction to Moral Philosophy* (Lanham, MD: Rowan and Littlefield, 2003); R. G. Frey, "Why Animals Lack Beliefs and Desires," in *Animal Rights and Human Obligations*, ed. Tom Regan and Peter Singer, 2nd ed. (Englewood Cliffs, NJ: Prentice Hall, 1989), 39–42; Alan White, "Why Animals Can't Have Rights," in *Animal Rights and Human Obligations*, ed. Tom Regan and Peter Singer, 2nd ed. (Englewood Cliffs, NJ: Prentice Hall, 1989), 119–121; Bob Torres, *Making a Killing: The Political Economy of Animal Rights* (Oakland, CA: AK Press, 2007); and Best, *The Politics of Total Liberation*. For an overview of the distinctions among modern animal rights advocates and liberationists, see Gary L. Francione, *Rain without Thunder: The Ideology of the Animal Rights Movement* (Philadelphia: Temple University Press, 1996). Prominent work in the field of cognitive ethology, or animal awareness of consciousness and intentions, begins with Donald Griffin, *Question of Animal Awareness: Evolutionary Continuity of Mental Experience* (New York: Rockefeller University Press, 1976); and Griffin, *Animal Minds* (Chicago: University of Chicago Press, 1992).

On the emotional lives of animals, see Jeffrey Moussaieff Masson and Susan McCarthy, *When Elephants Weep: The Emotional Lives of Animals* (New York: Delacorte, 1995). On the legal rights of animals, see Steven Wise, *Rattling the Cage: Toward the Legal Rights of Animals* and *Drawing the Line: Science and the Case for Animal Rights*, esp. chapter 12.

16. On the benefits of livestock farming and its connections to sustainability, see D. K. Beebe, "Animal Agriculture: How Can It Be Sustainable in the Future?," in *Sustainable Animal Agriculture*, ed. Ermlas Kebreab (Wallingford, Oxfordshire, UK: CABI, 2013), 284–311. On universal vegetarianism and veganism, see Tony Milligan, *Beyond Animal Rights: Food, Pets and Ethics* (London: Continuum, 2010), 92–104. On sustainable beef production, see Nicolette Hahn Niman, *Defending Beef: The Case for Sustainable Beef Production* (White River Junction, VT: Chelsea Green, 2014). A more anecdotal characterization of sustainable hog production can be found in Estabrook, *Pig Tales*, section 3.

17. Anthony J. McMichael and Ainslie J. Butler, "Environmentally Sustainable and Equitable Meat Consumption in a Climate Change World," in *The Meat Crisis: Developing More Sustainable Production and Consumption*, ed. Joyce D'Silva and John Webster (London: Earthscan, 2010), 175; Jonathan Watts, *When a Billion Chinese Jump: How China Will Save Mankind—Or Destroy It* (New York: Scribner, 2010), 294–317; Rohana et al., "Japan for Sustainability in Health through a New Movement of Food and Nutrition Education 'Shokuiku,'" 21–28; Poritosh Roy, Takahiro Orikasa, Manasiden Thammawong, Nobutaka Nakamura, Qingyi Xu, and Takeo Shiina, "Life Cycle of Meats: An Opportunity to Abate the Greenhouse Gas Emission from Meat Industry in Japan," *Journal of Environmental Management* 93 (2012): 218–224; and Tudge, "How to Raise Livestock—And How Not To," 14–15; and Tudge, *Feeding People Is Easy* (Pari, IT: Pari, 2007), esp. 70–74. For important perspectives on the type of diet recommended by Tudge, see Dan Barber, *The Third Plate: Field Notes on the Future of Food* (New York: Penguin, 2014).

18. Beebe, "Animal Agriculture," 298–302; and National Research Council of the National Academies, Committee on Twenty-First Century Systems Agriculture, Board on Agriculture and Natural Resources, Division on Earth and Life Studies, *Toward Sustainable Agricultural Systems in the 21st Century* (Washington, DC: National Academies Press, 2010), 4–5, 10, 13, 15, 22–23.

19. Salt, *Animals' Rights*, 96.

# Index